INSIDE OUT

INSIDE OUT

RETHINKING TRADITIONAL SAFETY MANAGEMENT PARADIGMS

First Edition

Larry Wilson and Gary A. Higbee, EMBA, CSP

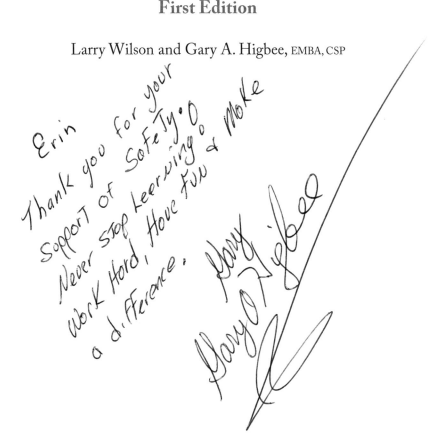

Erin
Thank you for your
Support of Safety. O
Never stop Learning.
work Hord, Have Fun & Moke
a difference.
Gary
Gary O. Higbee

Inside Out, Rethinking Traditional Safety Management Paradigms

Published in Canada by
Electrolab Limited
335 University Avenue
Belleville, ON K8N 5A5

Second Printing February, 2013

Editor: Jerry Laws
Marketing Manager: Ray Prest
Cover Design: Roseanne Carrington
Interior Design: Jillian Bower, Rachael Daniels
Printer: Performance Printing

Library and Archives Canada Cataloguing in Publication

Wilson, Larry, 1956-
    Inside out : rethinking traditional safety management paradigms / Larry Wilson and Gary A. Higbee. -- 1st ed.

Includes index.
ISBN 978-0-9881577-0-5

    1. Industrial safety--Management. I. Higbee, Gary A., 1947-
II. Title.

T55.W56 2012   363.11   C2012-905414-3

This book is dedicated to Donna Wilson and Jan Higbee for their support and patience during the writing of this book, and more significantly, throughout the ongoing development of SafeStart®, which now spans more than two decades. Their amazing willingness to pretend that our numerous vacations together were actually vacations, and not just facades for their husbands' endless need to collaborate on their work, is truly appreciated.

TABLE OF CONTENTS

Part 3—"Inside Out"—Reengineering the Workplace and Community for Safety

LIST OF FIGURES

LIST OF FIGURES

FOREWARD

OH&S already had published one of Larry Wilson's articles, "Better Methods, Better Results," when I came on board. The next article he submitted had a very controversial title. It was "The 'Dumb Worker': A New Perspective." Because it was new to me and intriguing, we published it in October 2002. Our magazine has published more than a dozen articles written by Larry since then.

In the past ten years, I've seen SafeStart grow from a very modest customer base to being used by more than two million people in 50 countries. Larry asked me to edit this book because I am familiar with his work and his writing. He told me up front that its format was somewhat unusual and said what he was trying to do was allow readers to experience what Gary went through as a safety professional.

While it is an unorthodox format, it works. You hear Gary's voice—and Larry's, too. Through the book, they reward readers not just by explaining how they came to understand what really causes most injuries and how to prevent them, but also by sharing practical steps and information required to implement this kind of change, and thereby significantly reduce injuries.

Jerry Laws
Editor, *Occupational Health & Safety*
Dallas, Tex.
April 9, 2012

PREFACE

As we all know, "the best-laid plans of mice and men often go awry."[1]

Well, I had a plan for writing this book: I would get Gary Higbee and his wife to come to Whistler,[2] and I would interview him—with a digital recorder—so safety professionals could hear it from his perspective, from one of their own (probably one of the most highly decorated one of their own); they could hear how and why his perspective changed, instead of just hearing it from me. Halfway through, however, he sort of turned the tables on me, and I found myself being interviewed. That really wasn't part of the plan—to tell everyone the story behind SafeStart. It just worked out that way.

So, in the end, you get to hear both perspectives. You also will find out how to apply these concepts and techniques to enhance near miss reporting, accident/incident investigation, job safety analysis, compliance training and your behavior-based safety process (if you have one).

Most importantly, you will see how these concepts and techniques can be applied to prevent motor vehicle incidents caused by human error and accidental injuries off the job.

Larry Wilson
larry@safestart.com
Whistler, B.C.
April 9, 2012

Beware of a man who invites you to go fishing but doesn't own a rod and reel.

Gary A. Higbee
gary@safestart.com
Johnston, Iowa
April 9, 2012

---

[1] This saying is adapted from a line in "To a Mouse," a poem by Robert Burns.

[2] Whistler is famous for skiing (2010 Winter Olympics) but started as a fishing camp because of how good the fishing was in the nearby lakes and streams.

ACKNOWLEDGEMENTS

We recognize the individual and collective efforts of the following people for their contributions in producing this book:

Jerry Laws for editing and feedback

Sheree Carlisle for administrative duties and proof reading

David Threlfall for content review

Jillian Bower, Rachael Daniels, Roseanne Carrington and Ray Prest for design and production

LW/GAH

PART 1

"Outside In"—

A Brief Look at the History of

Safety Management and

Extrinsic Control

CHAPTER 1—*Introduction*

Gary is sitting down at the dining room table. I turn on the digital recorder and the red light comes on, the signal that we are recording. Gary looks at the red light, pauses and then looks out the window. Gary loves to fish. "So here we are," he says. "We're in British Columbia. We're in Whistler. It's beautiful: there are mountains and a wonderful lake out here. I see a few people water skiing and one guy fishing—and you want to write a book about safety. You know, that's not the most exciting thing in the world. Why do you want to write a book about safety?"

"Well, it's not really just about safety," I said. "It's about getting people to rethink a lot of their paradigms in terms of safety and how safety is managed. A book gives you one of the best vehicles to explain something in detail. You need to be able to get into a fair bit of depth before people are going to even be willing to consider rethinking their paradigms. So that's why, I guess...." My voice trailed off.

"Okay," he said. There was another pause as he watched the guy who was fishing at the far end of the lake. I know how much Gary loves to fish, and I was starting to worry he'd suggest we rent a boat and do this out on the lake. Not that I've got anything against fishing, but I was worried we might run out of time. However, as it was, he turned his chair around so he wouldn't have to look at the guy and then focused back on the digital recorder. "So the whole idea," he continued, "is to help other people improve what they're already doing with injury prevention. Is that what you're after?"

"Yeah," I said, "but it's more than just injury prevention. It's really about reducing human error, which is a bit of a paradigm shift for a lot of people."

"Okay," he said. "So if I buy into this noble cause, does this mean I have to write half the book?"

"I don't know about half the book, but in terms of why *you*…well, I thought you could probably present the historical perspective and the traditional perspective," I said.

"Is that because you think I was alive back then when all this started?"

"No," I said, laughing, "I know you haven't been around *that* long, but you have been in the field for 40 years, which is longer than most. And, although I can't believe I'm quoting my dad, 'History gives you perspective. It helps you to see where we've come from and why we are where we're at.' But the other thing is that when you and I are presenting these concepts and techniques at a one-day or at a two-day workshop, we get enough time to explain everything in enough detail that people really get a chance to examine their paradigms. When they come in, they believe in what they've come to believe in, and people don't just naturally turn 180 degrees in five minutes. They don't even turn 90 degrees in five minutes. They need time to think about things. Not everybody can go to a one-day or a two-day workshop, but they can read a book."

"So you want me to talk about the history of safety," Gary said. He paused for a moment to collect his thoughts. "Well, it's been around for a long time. We even saw things in medieval times where they made some attempts to provide people with a little bit safer environment, just so they could protect the human resource. But it wasn't widespread, and it wasn't very pretty. In fact, if you go back a hundred or a hundred and fifty years, even then, it wasn't a very pretty picture. We didn't see much in the way of safety—or safety efforts—early on. But it was an interesting time."

"The industrial revolution in Europe was different than in the United States. Europe's industrial revolution was still largely based on human labor, while the U.S. industrial revolution, which occurred a little later than Europe's did, was much more mechanized. The injury rates, in industries where injury statistics were kept, were much higher in the U.S. than Europe. This can be partly explained because of the greater use of mechanization in the United States. Workers, many of whom were European immigrants, who flocked to the new factories and railroads, accepted the risks of their new jobs. They just needed the work, and besides, the wages were higher than what they would have ever been able to get in Europe."

I knew that there weren't any workplace safety records being kept back in the 1500s or 1600s, but I wanted to know if there was anything in terms of data and statistics for injury rates or fatality rates during the Industrial Revolution. I asked Gary what, if anything, they had for statistics back then.

"We don't have much data, that's correct," he said. "Nobody was really required to keep those kinds of records, but there's some interesting information that did get collected. It wasn't like people were ignoring safety. For instance, a man by the name of Mark Aldrich[1] gathered some data that he found from 1890 through

---

[1] Mark Aldrich, "History of Workplace Safety in the United States, 1880-1970," <http://eh.net/encyclopedia/article/aldrich.safety.workplace.us.html>, accessed April 9, 2012. This article is based on Aldrich's book *Safety First: Technology, Labor and Business in the Building of American Work Safety, 1870-1939*, John Hopkins University Press, Baltimore, 1997.

1904. Between 1890 and 1894, for every thousand workers in the mining industry in Great Britain, 1.6 of them turned out to be fatalities. In other words, 1.6 out of every thousand workers was killed. But in the U.S. the statistics for the coal mining industry were separated into bituminous and anthracite mining, which is kind of interesting because the process is about the same. The fatality rate for bituminous coal was 2.5 workers per thousand, and anthracite coal was almost 3.3 workers. Now, from 1900 to 1904 in Great Britain, that number was roughly 1.2, and the bituminous group was now about 3.5. So it appeared things were getting worse in North America, and they're getting better in Britain. Safety for anthracite coal was a little bit better, but it wasn't significantly improved over what we saw in 1894. So we saw the mining industry tracking injuries that far back."

"That means if you averaged Britain and North America, you're a little over two deaths per thousand workers per year?" I asked.

"Yeah, and for the time it was a fairly acceptable number," Gary said. "That was the way it was. But it was higher in the U.S. than it was in Great Britain."

"Did anybody else keep records? Were any other industries keeping statistics?"

"The railroad industry," Gary replied, "which, by the way, was one of the pioneers in terms of safety systems, also kept track of fatality rates. Their records were a little more comprehensive in that they kept track of fatalities for various work activities like coupling and braking. In Britain there was a reduction of injuries when comparing 1889 to 1895 to 1901: 4.26 trainmen per thousand in 1889, 3.22 in 1895 and then 2.21 in 1901. The American railroad workers during those same years, though, were running much higher. They were 8.52 for all causes, and then 6.45, and then back up again to 7.35. So there was a big difference between the fatality rates when you compare Britain and North America."

"Was there anybody offering opinions, speculation, on why that was?" I asked.

"Not really. Like I said, no one was required to keep these records. The companies who kept them did so for internal purposes, and didn't share the information with their competitors. We know that there was some worker dissatisfaction with the risks involved in mining and railroading. Worker safety started becoming much more important between 1900 and World War I. A lot of different groups were saying, 'Hey, this safety thing is a problem.' Unions were becoming more active and demanded a safer workplace. As industrial accidents with their fatalities and serious injuries grabbed the headlines, government, business and community leaders started campaigning for better working conditions. In 1893 Congress addressed railroad safety with the passage of the Safety Appliance Act and addressed concerns in the mining industry by establishing the Bureau of Mines in 1910. But I think that the most important thing in terms of substantial improvement was the workers' compensation law."

"And when was that, around 1900?"

"In 1908 Congress passed the Federal Employers Liability Act," he said. "That made it more expensive for companies to have accidents on the books. Although it was limited in scope, only covering railway workers who were involved in interstate commerce, it did open the door to some future legislation that would eventually apply to all industries. But what this law did is that it made employee error and intentional misconduct a defense that no longer could be used by the employers to deflect responsibility."

"Interesting. So they used to be able to claim employee error or intentional misconduct, and that was one of their defenses?"

"It was such a good defense that employers were very seldom paying any kind of real money, even for fatal injuries," Gary continued. "For slightly less than 50 percent of the fatal injuries, families were not paid anything at all. So what this law did was it made it harder for companies to use that claim. They used to say, 'The employees were at fault, therefore it's not our responsibility.' But now that they had to pay, the injuries were starting to cost companies a lot more money. The increase in cost per injury, at least in the railroad industry, went up tenfold. So now, companies in the railway industry had to pay at least some attention to injuries, and they also started to react to prevent similar injuries in the future."

So safety management started because of the money, at least for the railroad industry, I thought. And I guess that's not too surprising: Safety becomes a cost to the company, so management personnel—who are in charge of costs—start trying to control those costs. I wondered whether this is where the whole "outside-in" approach, or the idea of "extrinsic control," probably started, with the railroad industry back in the late 1800s.

"What about other industries?"

"They came along a little bit later," Gary said. "When they did, a movement was started. The movement to ensure that companies were held responsible for workplace injuries moved rapidly. In 1910, the state of New York passed a workers' compensation law that would have forced companies, all companies, to automatically compensate employees for workplace injuries. What this meant was that now the injured workers and their families did not have to take companies to court to get compensation. That was important because it took a lot of money to fight the large corporations in court, and their chances of winning weren't high. Unfortunately, the New York Supreme Court held the law to be unconstitutional."

"Wisconsin, in 1911, was the first state that successfully established a workers' compensation program that passed constitutional muster. Within one year, Wisconsin was joined by eight other states, and by 1920, forty-three states had some form of a workers' compensation law. By 1948, all 48 states had workers' compensation programs in place. Workers' compensation laws and other liability costs suddenly made workplace injuries an expensive proposition for many employers. What followed was a slow but steady improvement in workplace safety. Large firms in railroading, mining and manufacturing suddenly became more interested in safety. Manufacturing companies began to work to create safer equipment, and managers in many industries began getting tasked with identifying workplace dangers. In mining and construction, for instance, workers actually started to wear personal protective equipment, like safety glasses and hardhats."

"So these laws did a lot for improving safety, even if it was from an outside-in, extrinsic control perspective?" I asked.

"Yes," Gary said, "and although a lot of people don't know it, this idea of workers' compensation actually came from Europe. Way back in 1884, Chancellor Otto von Bismarck had initiated the first workers' compensation program in Germany. So, again, it was something borrowed from Europe."

"When did trade unions really start to become more influential in the area of workplace safety?"

"The big push for trade unions was the early 1900s," Gary said. "That's when unions really moved into being champions for safety. It wasn't like everybody was resisting safety. Everybody thought improving safety was a good idea."

"Okay," I said. "When did things like basic engineering safeguards start to come into play, where it would be commonplace to see a guard on a sprocket or a chain drive? When did any of that start to come into play?"

"Interestingly, as the workers' compensation laws came into being and started to change the playing field, what happened was that a whole lot of different things started happening all at once. The movement was fueled even more by one particular tragedy. On March 25, 1911, there was a fire at the Triangle Shirtwaist Factory in New York. The fire was caused by lack of adequate safety measures and took the lives of 146 workers, mostly young women. This event led to the creation of the United Association of Casualty Inspectors later that year. This group would later become the American Society of Safety Engineers."

"In the Midwest in 1912, there were some engineers, insurance company people, industrial people and even some government people who got together as a group for the sole purpose of promoting safety. That group eventually became the National Safety Council."

"As a result of all this activity, we started to see some very basic workplace safety improvements. Some of the first safety efforts included placing guards on machines and other equipment, and issuing personal protective equipment to workers. The cry for safer and better working conditions was getting louder. Now, there was still a fair bit of resistance to workers' comp, and there was actually a court case that went to the Supreme Court. But in 1917, the Supreme Court just stopped a lot of the controversy when they said the workers' compensation laws in the states were constitutional."

> I was pretty sure that the same kinds of things were probably happening all over, but I wanted to make sure. So I asked if this was happening everywhere or just in the U.S.

"I've been told that the ideas spread from Europe rapidly through the United States, and in some respects Canada. As a result, the previous concept of requiring workers or employees to sue for damages in court was just going out the window. Initially, there was some resistance to it. But, before we get hung up on the big bad companies abusing employees, we need to remember that many companies were already working pretty hard on safety, and some of them were doing quite well. DuPont, for example, was working on it during that time, and they were trying to control injuries long before workers' compensation laws came into effect. Unfortunately, these types of companies didn't appear to be the norm in industry. But my point is that it wasn't *all bad*. And workers' compensation did get some real supporters among many businessmen because, for one thing, it made the costs associated with employee workplace injuries more predictable. That was the one thing that businessmen really liked about it."

"Really?" I was somewhat surprised. "Wasn't it more expensive to pay for workers' compensation insurance instead of the odd lawsuit?"

"The thing is that, if a case went to court and the employee won, there was no ceiling—prior to workers' comp—on what the company might have to pay," Gary said. "Now, you're right, they didn't lose very many, but for the ones they lost, there was no set limit. So the idea of having some kind of restriction on the amount a case could get up to was appealing to them. You see, the workers' compensation laws really involved a series of compromises that centered around three areas: (1) no-fault insurance, (2) structured payments and (3) limited liability."

"Employees wanted no-fault insurance so that it didn't matter why or how the employee got hurt. Whether it was employee misconduct or not, they wanted to be covered by workers' compensation."

"Employers accepted their responsibility to take care of injured workers, but they needed some limits so they wouldn't go broke. That involved a structured way to calculate how much you were going to pay a week for an employee that was off work due to injury. Each state was a little bit different, but the amount was based on an employee's average income or a percentage of their average income so an employer's responsibility was capped. If an employee was off for a week, there was a maximum amount that they could get. And, if they had a permanent partial disability—in other words, they lost the use of their leg or their hand—they were also given a set amount. This structured approach of workers' compensation helped to limit the cost to employers."

"There was one final element that limited employer's liability. Workers' compensation became the employee's sole remedy. So, in exchange for no-fault insurance, employees gave up the right to sue, except in cases of blatant gross negligence. The employees also had to accept this structured payment. So, in spite of the compromises that each side made, employees and, in time, many of the businessmen and women decided that workers' compensation wasn't all that bad."

"Now what about the labor unions? Did they like this limited compensation and sole remedy thing?" I asked.

"At first, if you read the literature, they didn't like it at all. They thought it would mean that because there was a 'limited' liability, the employers wouldn't be spending a whole lot of time and effort on prevention. But, over time, they kind of got the idea that it might be okay. One of the people that led that was a man by the name of Samuel Gompers, the leader of the American Federation of Labor. He studied the effect of workers' compensation laws in Germany and was impressed at how it did just the opposite of what a lot of labor was afraid of: it stimulated business interest in safety. So his conclusion was that it was actually pretty good."

"Okay. So workers' compensation had a lot to do with companies starting to work on improving employee safety, didn't it?" I asked.

"Well, let's just say it really changed the playing field."

"In your opinion, would it be the main driver in terms of what improved workplace safety up to, say, 1970 and the creation of the Occupational Safety and Health Administration (OSHA)? Because the way I look at it, there had to be a lot of success—at least initially—with the outside-in approach (extrinsic control) for this kind of thinking to become so entrenched."

"There were some other things going on at the same time," Gary said, "but workers' compensation certainly opened the eyes—and the pocketbooks—of employers. But it wasn't the only thing. We saw some other activities that were really, really helpful. One was the American Standards Association. These folks got together with the people developing the voluntary standards in different industries, and they eventually became the American National Standards Institute, ANSI. So now we at least had some standards, but they were still voluntary."

"Another thing was the push toward comprehensive federal regulations. It began in earnest when Frances Perkins, who was the Secretary of Labor back in 1933, pushed really, really hard to get some kind of a federal occupational safety and health law. Now, it didn't work. She didn't get it done, but her efforts started to open the doors: the Fair Labor Standards Act (1938), Federal Coal Mine Safety Act (1952), Federal Metal and Nonmetallic Mine Safety Act of 1966, the Department of Transportation (1966) and the National Transportation Safety Board (1966)."

"You can see that the emphasis on safety and federal regulations is increasing. By 1968, President Lyndon Johnson tried again what Frances Perkins did by trying to get a Federal Occupational Safety and Health Act. Well, it didn't happen in 1968 either, but in 1969 we did get the Construction Safety Act. Also in 1969, an interesting thing happened that affected the practitioners of safety, those people who are safety managers or directors. The Board of Certified Safety Professionals was established, and it gave us some certification opportunities for practitioners in the safety profession. It raised the credibility of the profession and, because they really worked hard at trying to find new systems and new ways to protect employees, we saw the Board of Certified

Safety Professionals really become effective."

"Then, of course, in 1970 OSHA was established and also the National Environmental Protection Act, which created the EPA. So by 1970, we had a lot of things in play. However, you have to remember that in 1970, OSHA was just a baby. It wasn't even until May of 1971 that we actually saw the first OSHA standards. But that at least gives us a kind of a baseline in terms of where the compliance side of safety began."

> I was curious. It seems to me that today's safety professionals seem to care more about OSHA—getting inspected and cited—than they do about workers' compensation premiums.

"When, in your opinion," I asked Gary, "did OSHA start to make its presence felt? And when, if ever, do you think that OSHA sort of superseded workers' comp in terms of employer motivation?"

"By 1972, OSHA was well on its way. And, of course, I was in the business then. We really were quite concerned about what OSHA was going to do. We didn't know what was going to happen. It's like any change that you go through in life. First thing, you get the shock, and then you have to figure out what the rules are and how to abide by the rules. Eventually those rules become the new norm. We were really struggling. But we did get some really good guidance in a few areas—namely, personal protective equipment and machine guarding. Those two gave us a handle on some things that were pretty well structured. The fear of compliance didn't necessarily put workers' compensation on the back burner. By 1974 or 1975, they were working hand in hand: companies were trying to control workers' comp costs *and* were trying to avoid penalties for non-compliance with OSHA standards. The interesting thing—and, unfortunately, the worst thing—is that the safety practitioner was working on compliance, and the risk manager was working on control of workers' compensation costs. Their skill sets were completely different, and their focus was completely different."

"And was what they were accountable for completely different too?" I asked him.

"Oh, absolutely. All the safety professional was accountable for was to make sure we were in compliance. The risk manager was worried more about how much the insurance was going to cost and how they were going to negotiate new insurance contracts. Even back then, that was a problem. Instead of working together, they were separate entities. In fact, often the risk manager was a corporate position, and the safety practitioner was in the line organization."

"How much, if any, has that changed up to now?" I asked.

"Well, we've seen some changes, but I don't think they're very close yet. They are in some organizations, but not very many."

I had to agree. I certainly hadn't seen it in terms of safety practitioners being aware of how much financial exposure the company had from off-the-job injuries and injuries from preventable vehicle accidents. Even now, it's rare to see the two working hand in hand at most companies.

"What was interesting," Gary continued, "was that the safety practitioners were really working hard on compliance, but as they got better with compliance, the problem was that they didn't really see their injury rates going down as fast as they would have liked. Because the compliance piece couldn't answer why somebody slipped or tripped, or why an automobile accident happened while they were driving a company car. There were so many things outside of the control that legal compliance provided that I, for one, was struggling."

CHAPTER 2—*The Five Stages to World-Class Safety*

So things were really bad 150 years ago from a worker's safety point of view, and pretty bad 100 years ago. Then they got progressively better, in terms of injury rates, because of workers' compensation, and, after 1970, because of workers' compensation and OSHA. Now, I at least knew how the whole outside-in approach started. And it made sense: Guarding things and guarding people are very efficient, at least up to a certain point they are anyhow. But I was still looking for perspective. My guess was that there were some really good companies. There were probably fewer (unfortunately) great companies, a lot of middle of the road, and a lot of not so good ones.

When I asked Gary about this, he said, "You're right, companies were all over the place. What I used to see was what I eventually started calling the five stages to world-class safety. For years I used that as a guide to help me place people or organizations where I thought they were in terms of safety performance: how sophisticated their management systems were, where they were at with behavioral safety systems and what they needed to do to get better. If you look at the diagram (see Figure 1), you'll see

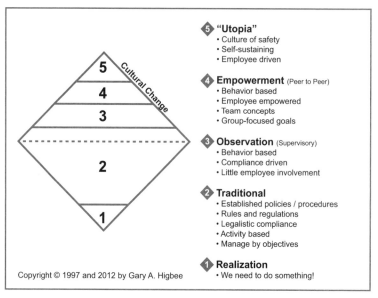

Figure 1
Five Stages to World-Class Safety

that the area of the diamond for Stage 1 is much, much smaller than the area in the diamond for Stage 2. What that means is that I saw way more companies in Stage 2 than in Stage 1. It's just a visual representation of where I thought companies were then. Frankly, I think it's still valid in terms of where industry is today."

"So, the areas simply represent the approximate number of companies in each of these five stages?" I asked.

"Yes," Gary said, "Stage 1 is what I call the realization stage. These are the companies that are hemorrhaging in the area of safety. They do not have safety systems in place; their workers' compensation costs are out of control. In fact, they might even have trouble hiring employees because nobody wants to go to work there. They have never kept up with the safety regulations; they get by with as little as they can. But, at some point, they finally come to the realization, 'Hey, this isn't working. We're going to have to do something here.' So they realize that they've got to improve. Typically what happens is they will then quickly move into Stage 2."

"Stage 2 is where you have the traditional safety system. It's your policies and procedures, all your rules and regulations. Companies who move up to the bottom area of Stage 2 are in the process of establishing those policies and procedures or improving the ones they have. They're looking closely at their rules and regulations. They're asking, 'Which ones make sense?' and 'Are they written right?' This stage is primarily based on legalistic compliance: meeting the regulations, whether it's OSHA or any other regulator. The major safety activities are geared toward meeting those compliance requirements, whether it is training or inspections or audits or whatever it might be. They manage them in some kind of a system, and they have objectives for improvement. So essentially, they manage by objectives. You don't jump from the top of Stage 1 right into the middle of Stage 2. It's a slow migration or progression that you go through."

I'm looking at Stage 2 on the diagram. "Okay, is this where you'd do one of your audits to see if a company is in compliance with the regulations and their own written safety programs?"

"Yes, this is the stage where a company would sometimes arrange for an independent third party to come in and audit or inspect what's going on. I've done quite a few audits myself. What we're really doing is looking at all the safety systems: Are their written programs proper? Are they doing the right job? Are they meeting the standards? After the systems are in place and all the procedures are there, the hope is that what they actually do is what they say they do in their programs. As you can imagine when you first go out on the shop floor, quite often there's a big difference."

"So as you keep making efforts in terms of improving your 'written' compliance, you move up to that dotted blue line (see Figure 2)."

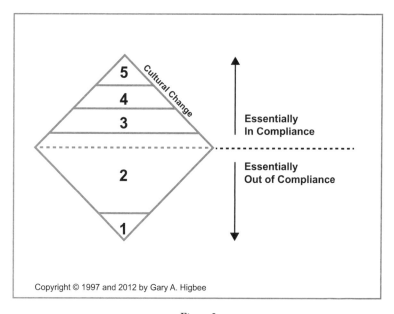

Copyright © 1997 and 2012 by Gary A. Higbee

Figure 2
Five Stages to World-Class Safety
The Blue Dotted Line

"When you get above that dotted blue line, you are essentially meeting all your compliance or regulation requirements. Now, the reason I've got a little bit of space between that dotted line and the line at the bottom of Stage 3 is that this area allows for some employee participation. Below the dotted line, there's not much employee involvement with these companies. But above the dotted line, there are companies that have safety teams or safety committees, and they get their wage employees involved with inspections for unsafe conditions. I used to think that when you got to the line right at the top of Stage 2, you were world-class."

"Well," I said, "that kind of thinking was very common at the time, and I suppose to a certain extent it still is. I mean there are still a lot of people out there who think that if they could get their plant totally up to legal compliance standards, they were there—they'd made it."

"If you didn't fear the regulator coming in," Gary said, "you felt you had a pretty good handle on safety. The assumption was that if you met the compliance standards, you would have very few injuries. In other words, we thought that improved compliance would directly relate to further injury reductions. And it does, up to a point. But I had an incident in one of my facilities that changed my perspective on all that. I had a facility that was, I'll say, the 'darling of OSHA.' They used that facility as a model of what a compliant facility should be. When they had new inspectors, they would come to that facility and say, 'Hey, can we come in and take a look?' They'd come in, and they would find a few minor things, but they were essentially telling the new OSHA inspectors that this is basically the way it's supposed to be. And so I thought we were there, I really did. I had no fear of the regulator. Life was pretty good."

"If they were bringing inspectors into your plant to show the new inspectors the way it's supposed to be, I wouldn't think you'd fear the regulators," I interjected.

"Not at all. Even though they would find some things from time to time while they were in there, they were solved easily, so there weren't any major issues. Everything seemed to be going fine. Then, as I said before, I had an incident in the facility that resulted in a death. Of course, when something like that happens, you ask yourself a lot of questions because you're really, really frustrated."

I had heard him tell this story before. It wasn't a happy story. But now I was (finally) getting it put into perspective—in terms of Gary's journey.

All I said was, "You knew the person very well, too."

"I knew the person very, *very* well—I don't want to say much more than that. But yes, I knew the person very well. In fact, we did activities outside of work on occasion."

Again, all I could say was, "That must have made it even harder."

"Well, it certainly made it worse. But *any* death is a bad situation. I couldn't understand why it happened in that facility. My assumption all along was that if we were compliant, everything ought to be okay. This incident had seven or eight contributing factors to it, just like many incidents do. And if any one of them hadn't happened during a particular time frame, then the others wouldn't have happened either. But they did, and I was tremendously frustrated. So within days, I started going out and taking what now are called 'behavioral safety observations.'"

"So was that, if you will, the point in time when you realized, 'Okay, there was more to safety than just legal compliance'? Was that the turning point for you?" I asked.

"Yes, I finally realized that compliance wasn't enough. But what I didn't know, though, was *what was enough*. That's what I was searching for. And I was really concerned that I didn't know what else I could have done to have prevented that fatality."

As I listened to Gary's words, "what else I could have done," I realized that, at least at that point in time, he was still going down the wrong road. What he should have been asking was, "What concepts and techniques could I have given the employee?" And then say, "What could I have done to help him understand and develop those safety skills?" That would have been, in retrospect, a much better way to approach this.

Meanwhile, Gary continued. "So then I started spending more time on the floor, even though I already spent a lot of time on the floor. But now my time was more structured. I was watching very specifically what people were doing and what circumstances they were in. And, frankly, I was appalled. I saw truck trailers without the wheels chocked. Forklift operators who were driving on and off to load them, even though we had a procedure that said you're supposed to have the wheels chocked, and it's supposed to be double-checked. I found people driving forklifts without having done the daily inspection. And worse, I found out that they were doing the inspections at the end of the week and turning all five of them in at the same time. So I was really frustrated by all that."

I could sense he was getting a bit worked up, so I tried to lighten the conversation a bit. I told Gary that I once was at a place where they had the inspection reports filled out a month in advance, just in case they showed up late one morning. Because if something happened when they were on the fork truck and they hadn't filled out the report, they were in a lot of trouble. Filling them out a month in advance, just in case they didn't have time to get the inspection done, meant they never had to worry about it. My hope was that Gary would realize this kind of stuff went on everywhere.

Gary just shook his head. "So, like I was saying, what was happening was, even though we had all the procedures written down and they looked fine, and even if the regulator came in and he thought everything was fine, it really wasn't because the procedures weren't being followed."

"So the filing cabinet looked good," I said.

"The filing cabinet looked *very* good. And the reports looked good."

"But when you actually went out and looked on the floor, the difference between what you were supposed to see and what you actually saw, like you said, was shocking?"

"It was. I don't know how or why I was so naive, but I was absolutely shocked."

I told Gary that I wasn't surprised by the gap between

what was written down in the safety office and what was actually being practiced in the plant. At the same time he would have been going through this, I was teaching a supervisory observation process at companies all over Canada, in all different types of industries. One of the things we would do in the afternoon is pair employees up in teams of two and send them out to different areas of the plant. In most cases, they would come back with that kind of "I wouldn't have believed it if I hadn't seen it with my own eyes" look on their faces. They were really, really surprised at what they saw in terms of just basic compliance, things they thought they had covered but didn't, whether it was personal protective equipment, machine guarding or even a critical procedure like lockout/tagout. There was a lot of surprise and shock and, in some cases, embarrassment. But I don't think they would have believed it until they saw it with their own eyes.

"Well, there's a reason for that," Gary said. "It's because most managers think that injuries are the only measure. As a supervisor, if you don't have any injuries, then therefore, you must have a good safety program. So what I was trying to do was to find a different way to measure risk. We needed to measure risk, not merely injuries. The reason is simple: you don't get enough injuries to really give you a clear picture of the actual risk. When things are going well, you don't really have any data that tells you where your potential problems are, unless you start measuring the near misses and close calls that are on the bottom of a traditional risk pyramid (see Figure 3)."

So what Gary was doing, way before it became "trendy," was trying to get a handle on "leading indicators": those factors that change before the injury statistics do. If you don't have a way of measuring them, then there's not enough in terms of warning signs. But more important than measuring or quantifying leading indicators is

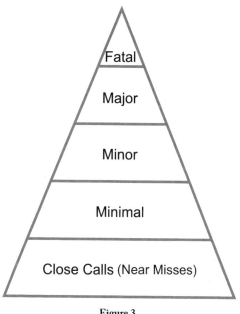

Figure 3
Injury Risk Pyramid

finding a way to improve them. The thing that I used to say to companies was that the measuring of the improvement isn't as important as the improvement itself. So if you believed the activity of going out and positively correcting deficiencies or positively reinforcing safe behavior is going to improve safety, then you get out there and you do it. You wait a while and usually, yes, you start to see things like the "percent safe" going up. But the reason it goes up is because people were out there making positive observations. Eventually you did see the injuries coming down, but there was usually another delay before that happened, too.

As I explained this to Gary, he said, "Well, I guess I actually walked a little bit of the same path, even though I didn't know what these observation processes were. But I started asking my supervisors to observe their employees periodically during the day doing certain work activities with a little checklist, so they knew what they were supposed to look for. The checklist contained things like following proper procedures, right tools for the job, wearing the proper PPE, that sort of thing."

"So this is the observation component you're talking about in Stage 3?" I asked (see Figure 4).

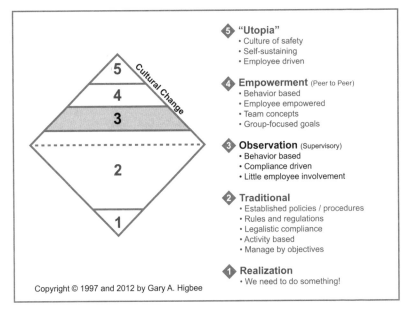

Copyright © 1997 and 2012 by Gary A. Higbee

Figure 4
Five Stages to World-Class Safety
Stage 3: Observation

"Yes. I migrated into Stage 3, and I thought, 'Okay, if we're really, really good at the traditional program, how are we going to find out if we have gaps in it? We're not going to find out if we have gaps by waiting for the injury to happen, that's not the way to approach it. People can make lots of mistakes, and we can have lots of risk and lots of issues out there that don't ever result in injuries, we'll never find out where we need to focus our attention on if we only look at injuries.' I wanted to see where our issues were, so I started having each of the supervisors, much to their disappointment...."

"Chagrin?" I interjected.

"'Chagrin' would be the right word. They didn't really like it, but they started filling out these observation checklists. Now, how well they were being filled out became an issue, and how accurately they were being filled out also became an issue. So I had to spend a lot of time sort of coaching them. But I was finding out, and they were starting to discover, some of the things that they should, quite frankly, have been doing all along. In fact, I believe if supervisors are doing the right thing, you don't really need an observation process. But many times they aren't, at least not always. You have to guide them and give them some structure so that you're at least getting observations. The idea was that you could build on your traditional safety program by making observations to see if there were gaps in the system and to use those observations to help quantify risk."

"You're observing for things in the traditional safety management system like personal protective equipment, following procedures, orderliness, standards, tools, equipment, that sort of thing?" I asked.

"Right. It was all still based on compliance, or what we thought our traditional safety program should be," Gary said.

"Now, when you were doing this, would you also be observing for and encouraging your supervisors to observe for ergonomic issues, or had you not quite gotten into that at this point in time?"

"We hadn't got into that. We were really still—well, I guess I was—still stuck with compliance."

"Well, it really wasn't until post-1990 that ergonomic risk factors started making it into the behavioral safety world," I said, "whether it was a supervisory-driven program or a peer-to-peer program." That thought led me to my next question, "Stage 4 of your five stages says '*Peer to Peer.*' So is this simply changing the observation process from being a supervisory-driven process to a peer-to-peer process?"

"Not entirely," Gary replied. "Stage 4 does involve peer-to-peer observations, but I didn't stop the supervisor-to-employee process that started in Stage 3. What I was finding out was that the supervisors were learning as much by observing employees as the employees were while being observed. I thought this was interesting: Some of the employees actually came to me with an idea of their own. In those areas where you had a lot of people and only one supervisor or on the off-shifts where there wasn't any supervision, they thought it would be a good idea if some of them were trained to do observations. So we started to train the employees to do what later became known as 'peer-to-peer' observations. We also had to work through some potential discipline issues and the apprehension some employees had because they were worried they might be 'ratting' on their workmates."

As I listened to Gary, I had to agree. There were certainly a few issues or potential problems for companies moving from a supervisory-driven process to a peer-to-peer process, although (like Gary) I never encouraged the supervisors to stop observing. Because as long as everybody keeps it positive and friendly, it doesn't matter who does the observing. I found the difficulty with the discipline issue was almost always with the supervisory observation process, because they had the authority to discipline people. But when you move into the employee-driven process, it was never much of a concern any more for the employees because the co-worker really didn't have the authority or the wherewithal to discipline them. I did find that it wasn't always as easy to move into peer to peer as it should have been. Even if the supervisory observation process went well, the employees didn't always pick up the ball and run with it. But most of them did.

"So," Gary continued, "I thought, well, if this observation process for supervisors was helping us find gaps in the safety system, then maybe it would be beneficial to actually have peers observing peers. The idea was that if the peers were observing peers, then both the observer and the person being observed would learn a lot about their behavior and how it related to the traditional safety program. So we were starting to get peer-to-peer involvement. I thought the more you got people involved in this process, that eventually they would do it in real time. In other words, they would actually do the observations when they really weren't doing observations, and everybody would be looking out for each other. If we get to that point—where everybody's looking out for each other—and we know what to look out for, what to observe for, how to look for gaps in the system, and what to look for in our behavior as we go about our business, then we ought to have a very efficient, safe operation with supervisors and employees all looking out for each other. That would get us to Stage 5, which I considered at the time to be 'Utopia.' What I thought would happen in Stage 5 was that since we were so used to doing peer-to-peer observations and so used to doing supervisor-to-employee observations, we would all be looking out for each other in real time. In other words, you wouldn't have to consciously schedule an observation, you would observe almost continually."

"When I really thought we'd reached Stage 5 was when I got a phone call one day from a person that was doing asbestos removal. This was a big project in a power plant that we had where two teams of two were working about 40 feet (12 meters) above the floor in lift equipment using the glove bag method to remove asbestos. The teams were using fall protection, and had their disposable suits and respirators on. It was hot work, probably 140 degrees Fahrenheit (60°C) up there because the power plant was still working. The work area was all enclosed and had negative pressure. Two would work together on one lift, and the other two would work together on the other lift. However, the teams could only work for maybe 45 minutes at the most before the heat would get to them. Then we'd have to send in two fresh teams."

"So here's the situation. We had two people on one lift 40 feet (12 meters) in the air, correct PPE on, doing a good job, working on trying to get as much asbestos off of this pipe as they could. They looked over at the other two guys that were working probably 30 feet (9 meters) away on the same pipe, up there 40 feet in the air, and saw that both were wearing their disposable suits and respirators, but only one of them was using fall protection. When the first two saw that one of the other two didn't have fall protection on, they stopped working, lowered their lift, went over, motioned for the guys to come down and said, 'Hey, you don't have your fall protection on.' The guy said, 'Yeah, I forgot it when I came in, but I knew if I went out to get it, it'd take me ten minutes to clear out and ten minutes to come back in, and I wouldn't have much time left to work. So I thought just this one time I'd go ahead and work without it.' They said, 'No, no, we've got another one.' They found one for him within the enclosure. He put on the fall protection, and they finished their work."

"That afternoon when the guy who forgot his fall protection was finished, he gave me a call. So here's the guy who was removing asbestos up 40 feet in the air without his fall protection on, calling the safety manager to tell him what happened. What he wanted to tell me was that, 'These observations we've been doing are working. They're paying off because there's no supervisor or union official in there; nobody is in there except those of us that are working because we're the only ones who could be in that enclosure. Those guys thought enough of me to come down in all that heat, and come over and make sure I had my fall protection equipment. It slowed their work down, and in fact, they ended up working a few minutes longer in that oppressive heat.'"

"After I heard him tell me the story," Gary continued, "I thought, we got it! We're finally at Stage 5. We are getting peer-to-peer observations, people looking out for each other. They're not formal observations; they're informal. They're happening in real time—that's what I thought 'Utopia' was."

I listened to this and smiled as Gary finished up explaining his five stages. His description of "Utopia" reminded me of the top tier of a diagram I used in the past to show the three stages of behavior-based safety (see Figure 5).

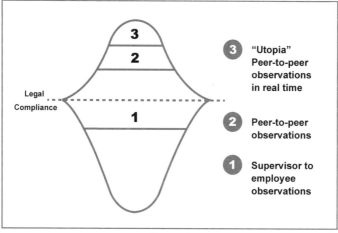

Figure 5
Three Stages to Behavior-Based Safety

"Hey," I said. "Remember that diagram I showed at the Dallas meeting?"

That was the first time I met Gary. We were both invited to attend a safety advisory board meeting in Dallas. The company sponsoring it was interested in finding out what people in the EH&S arena were thinking about the future.

"How could I forget that?" Gary responded. "Here I thought I had done a lot of work and really had something with the five stages I'd developed. Then here's this Canadian explaining to me about the top half of *my* diagram. I couldn't believe it. I wondered where in the heck you got it—thought somehow someone stole it."

"Well, I can assure you I didn't," I replied. "I think you and me, and a lot of other people were simply trying to graphically represent the developmental stages of a high performing safety system." Thinking about how similar the diagrams were, I said, "I was surprised when I saw your five stages too. You'd obviously given it a lot of thought. I really hadn't thought about your Stage 1 and Stage 2. When I first saw your diagram, it reminded me of a diagram that the Workers' Compensation Board of British Columbia had (see Figure 6)."

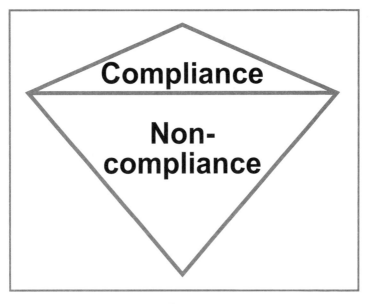

Figure 6
BC Workers' Compensation Board
Compliance/Non-Compliance

"Their shape was more like a diamond, and they were only trying to represent the relative number of companies in compliance and out of compliance. They were a lot less optimistic about the number of companies that were above legal compliance than you were. Although they didn't have formal stages, they did talk about where various industries like forest products, pulp and paper, logging, construction and petrochemicals were in relation to each other."

"You know, I'm glad we took the time to talk during our lunch break that day," Gary said. "I'd had a hunch that Stage 5 wasn't enough. I was wondering what else was out there. You gave me a lot to think about."

CHAPTER 3—*Persistent Problems*

Gary continued to reflect on that day in Dallas so many years ago. "The problem, or the disappointment, I had was that after we went through those five stages at work, I thought that was the end, but it wasn't. We still had a lot of issues that weren't being dealt with: sprains and strains, and slips, trips, and falls. Even in Stage 5, we really had nothing that would help with preventable vehicle accidents. So all of these things were still out there, and we weren't able to take care of them. I'm wondering what the behavior-based safety (BBS) people were trying to do at this time? After all, they had to be seeing some of the same things."

"Well, I don't know as they were really trying to deal with these issues, either," I said. At the time Gary was referring to (the mid-1990s), I was working with pretty much all of the BBS companies in North America. Our company, being Canadian, had access to the Canadian market, and most of the BBS companies did not have a viable way to market their consulting services and training programs in Canada. They were looking for a distributor/broker, somebody to help them with that aspect of it, and we were a good fit. So by the mid '90s, with almost everybody who was doing BBS work, we ended up marketing their materials in Canada. But they weren't really focused on what was left once you got to Stage 5—the slips and falls, sprains and strains, cuts and contusions. Mostly, they just talked about the benefits of BBS in Stage 4. But the reality of it was that they really weren't talking about the problems that still existed, even if you could get to Stage 5 and you got peer-to-peer intervention in real time.

After I explained this to Gary, he said, "It wasn't like we didn't know there was a problem. These types of injuries were a high percentage of the recordables. We knew that we needed to do something about sprains and strains, so we hired ergonomists. They were working on force, position and repetition, the design of the job, and trying to figure out ways to get rid of some of these 'sprains and strain' type injuries. The engineers were continually designing machine guarding and other engineering controls. Not only did they design machine guarding, but I even worked with some who designed 'safety envelopes' where if we couldn't guard the machine, then we would guard the employee. If the employee were to move out of a certain area that was considered safe, then the system would shut down. The engineers were getting the big stuff like kill switches and dead man's switches. But even with all this effort, the safety pros still knew there was something else missing. They were putting up signs like 'Think Safely' and 'Safety Is a State of Mind.' So as a profession, we knew that there was something missing, but we were struggling because we really didn't know what to do about it."

"I saw some of that, too," I said. "I'm sure the safety professionals, having gone through incident investigation after incident investigation, started seeing the importance of 'inattention.' But the problems—sprains and strains, preventable vehicle accidents, cuts and contusions, and slips, trips and falls—were still there. I suppose within that same category you've also got all the fork truck accidents and the accidents with mobile pieces of equipment. So, these issues would have involved a significant number of injuries, even with a lot of those other hazards mitigated by the safety management system. As a result, companies were having a very hard time getting to zero injuries."

"We were taking care of a lot of the accidents and a lot of the injuries," Gary said, "but we were still having trouble with slips, trips and falls, cuts and contusions, big problems with preventable vehicle accidents, whether they involved cars or forklifts, and lots of sprains and strains. It wasn't like in Stage 5 we were hurting a high percentage of people, but obviously we were still missing something, and it was hard to figure out what that was. So when did you start figuring this out?" Gary asked. "Did you have a feeling that you knew what was missing at this point in time?"

"I had a hunch back around 1992, that it was really all about people moving," I said. "I think that was the big epiphany for me. You see, you could use engineering controls and machine guarding to keep the stuff from hitting people, but it was really difficult to keep them from bumping into things, or from moving into the path of something that was moving. Those potential incidents were always going to be unpredictable from an outside-in, extrinsic control point of view. So that was really it for me. It was just sort of a hunch at the time, something I was certainly trying to explore because several of the companies I was working with were experiencing a lot of frustration. They were getting their lost-time accidents down, but not always to zero. They still might have the odd back strain injury, which would cause some days off. Their injury rates for slips and falls, and cuts and contusions—injuries associated with people in motion—were still high, although they had come down slightly."

"Well," Gary replied, "like I said, we—safety professionals— were struggling a lot. When compliance or really, really good compliance or Stage 5 'Utopia' wasn't quite getting it done, we needed something else to help with these injuries. Along came some new technology, and a new profession actually came into being. Ergonomics became a hot item, and we started talking about ergonomic issues causing a high percentage of our problems. What they were trying to do was to design jobs and systems with some adjustability and flexibility to try and make it easier for any employee to perform the job. They were concentrating on how much force the employee had to exert, what position he or she might be in, and how often they had to repeat those activities."

"But most of the focus was on cumulative trauma disorders?" I asked (although it was more of a statement).

"Absolutely," Gary said. "The main focus was on trying to get the repetitive strain injuries down. So their focus really wasn't on acute trauma like broken bones, cuts and abrasions or anything like that; it was almost solely on repetitive strain. These injuries were starting to become a much higher percentage of the total injuries because, as you get better at compliance and you get better at the observation process, the acute injuries start to diminish as a percentage of your overall injuries. So ergonomics certainly came into its own. OSHA even made an attempt to promulgate a federal standard. That didn't quite work out the way some people planned. But that doesn't mean the problem went away, and it doesn't mean that we're still not struggling with those issues today. But things certainly got better in terms of preventing cumulative trauma disorder."

"Right," I said, "but the ergonomists were looking at redesigning equipment to fit the human body because the human body couldn't be redesigned. Human factors engineering was still in its infancy, and ergonomists weren't really looking at it from a human fallibility, or human error, point of view when they were designing things. It was more of, like you say, the vibration, the force, the position, the repetition, not necessarily the slip and fall or the bang your head or the falling asleep at the wheel."

"Correct," Gary continued. "We were using Snook Tables and engineering design tables to try and get as much of the force, position, repetition, vibration, or cold and hot out of the job. But we were looking more at things that were causing carpal tunnel syndrome, trigger finger, those types of injuries. At the same time, working separately, not in conjunction with the ergonomists, were engineers. The engineers were trying to get better at machine guarding, and better at safety systems. We saw the design of systems that included safety envelopes, a lot more light curtains and screens, and the ability to stop machinery in mid-stroke or actually shut it down. They were also working with robotic systems and laser systems. But they were working independently of the ergonomists. The last group that was working independently was the safety professionals. If they really got their traditional processes working and they had an observation process, then they were probably still struggling with 'inattention' because these accidents don't make sense to them. It's like people weren't looking at what they were doing, or certainly not thinking about what they were doing. So as mentioned before, we saw all kinds of attempts to get at the problems caused by inattention or human error. We'd see signs like 'Safety Is a State of Mind' or 'Think Safely' or 'Think of Safety.' The whole idea was to try and get people's level of safety awareness up."

I told Gary that I'd seen a banner years ago inside a chemical plant that said, "Never Give Your Mind an Additional Task at a Critical Moment." When I saw it, I thought there was a great deal of "face validity" in that statement: that anybody who saw it would agree that it had a lot of merit in terms of safety and preventing accidents. However, you wouldn't find anything like that in the safety management system that this company had. There wasn't anything to address awareness, attention, human error, visual focusing or any of that stuff at all. When a person would see a banner like that, they'd take it at face value and say, "Hey, that makes a lot of sense to me: If you're doing something that's potentially very dangerous, don't give your mind an additional task; keep it focused." Nobody would argue with you about that. And yet, like I said, it wasn't part of their system or anybody's system at the time.

CHAPTER 4—*Behavior-Based Safety and Human Error*

I told Gary that at that time, the behavior-based safety (BBS) process was primarily aimed at deliberate risk. It wasn't aimed at unintentional risk or human error; it was aimed at what people did deliberately. And the supervisory observation programs were mostly built around compliance: the personal protective equipment that the people should be wearing, proper tools and equipment, procedures, orderliness standards, those sorts of basic categories. Now, according to the ABC theory that BBS people taught, the activator was only so motivating. It was actually the consequences that motivated people. So by making these observations—providing they were positive—you would be adding a different consequence to a safe behavior than what people normally got, which was nothing. As a result, they would be more motivated to do it safely or follow the procedure in the future. So it wasn't really aimed at error or inattention.

The other thing that I noticed was that none of the behavior-based safety approaches seemed to have much impact off-the-job, including how people drove their cars. While companies were improving at work and were reducing their injuries at work, they really weren't doing much in terms of preventable vehicle accidents. Unfortunately, that's where a lot of the fatalities happen. Off-the-job incidents really weren't being impacted at all. You didn't have people going home and saying, "Honey, let me give you some feedback on your at-risk behavior cooking dinner last night...." It just wasn't happening. So on the behavior-based safety front, there really wasn't anybody who was looking at human error directly, and they weren't even looking at off-the-job safety. But I wanted to ask Gary whether there was a belief within the safety profession that with the addition of the ergonomics and the BBS processes, they actually had everything covered.

"Some safety professionals thought so," Gary answered, "but I don't know how many. Because there were still lots of injuries, mostly the kind you mentioned: slips, trips and falls, sprains and strains, and preventable vehicle accidents. There were still too many of those happening for a lot of us to believe we had everything covered. But some people did."

When I heard this, it reminded me of when I was at an auto assembly plant that had 4,500 people. They had 17 recordable accidents on the stairways—just on the stairways. That was too many to ignore. I also knew there were some folks who sort of stuck their heads in the sand and pretended nothing was missing and the system was sufficiently robust. I remembered the safety personnel at a nuclear power plant who really had bought into this idea: "The system is omnipotent, the system will protect us, and the system is all powerful." It almost had a tinge of religious fanaticism to it that really scared me because, after all, this was a nuclear power plant. On the other hand, they also made jokes about how Homer Simpson working at a nuclear power plant wasn't a fluke. They said somebody on the writing team had to have worked at a nuke to know what it was like. That wasn't comforting, either. But all kidding aside, there was a real belief that the system really was the all-powerful Oz, so to speak. So I asked Gary why so many safety professionals had accepted the concept of extrinsic control when they had to deal with so many injuries that were caused by inattention and human error.

"Well, you have to understand that the system was also a security blanket," Gary said. "As long as you were doing what the system said to do, and as long as you were meeting the compliance requirements, you couldn't get into too much trouble, no matter how many injuries you were having. As long as you were following the system, all would be well. It was only when you, as a safety professional, strayed outside of compliance that you took a fairly big risk because you were trying to do something different. So what you would find at this point in time, and even still today, is that almost everybody is mostly driven by the regulations and the standards. That's why a lot of companies are still stuck in Stage 2. There are companies in all five of the stages, but Stage 2 is still the largest. There are all kinds of companies that are relying on compliance as the answer. Since it's safe and secure, and they're not going to be questioned when they do anything inside of the compliance realm, it's a pretty safe place for them to be. So quite a lot of safety professionals probably still do think the system will work, and they probably think they just aren't doing enough or working the system quite well enough."

"Yeah," I said. "I think you're right. But you couldn't ignore all of the injuries like slips, trips and falls that were still happening and, like I said, the majority of the BBS people weren't even going in the human error direction. I remember back in 1999, Jerry Laws, the editor at OH&S magazine, saying, 'Nobody's talking about this but you.'"

"So, how did you figure this out?"

"Well," I said, "this whole human error thing started on a personal basis. I was staying in the middle of nowhere in western Alberta, and I had to get up in the night to go to the bathroom. I didn't want to really wake up, so I didn't want to turn the lights on. On the way back I needed to get my lip balm because my lips were dry. I knew sort of where I thought I'd put it on the night table, but I couldn't find it rummaging around in the dark. I also knew there was a drink on the table and I didn't want to knock the drink over, I had another lip balm stick in my gym bag, so I stepped over to get it. Now before I went to bed, I'd pushed my shoes out of the way so I wouldn't trip on them if I had to get up in the middle of the night. Unfortunately, the shoes were now right in my way. My foot stepped on the shoe sideways, my ankle rolled over and I started to fall. So reflexively, I put my hands out. They hit the coffee table. But when my left hand hit a shirt that was folded on the table, my left arm swung out and caused my head to hit the corner of the coffee table right above my right eye (see Figure 7). The cut was about half an inch (1.3 centimeters)

Figure 7
The Coffee Table Incident

long, but it was in the eyebrow. I probably would have had it stitched up if it was anywhere else, but as it was I just put some ice on it and went back to sleep. However, by the morning I had a huge black eye, almost closed, and the black had spread over to the other eye."

"So there I am next morning trying to teach a behavior-based safety class with two black eyes. At the morning coffee break, one of the guys came up to me and asked me what happened, and I told him. He looked at the BBS checklist, which was sort of the traditional stuff—the tools and equipment, the procedures, the personal protective equipment—and he said, 'How would any of this have prevented your injury?' I said, 'Well, it wouldn't.' You could kind of see the look of defeat on his face as he looked at the card. It was like he was thinking, 'Then what am I in this class for?' So I hurried to say, 'Well, yeah, but there's a lot of stuff on the checklist that would be good for people to do safely, right?' And he said, 'Yeah, right.' But then he said, 'But I did the *same* thing! I tripped on the coffee table leg,' he said, 'coming back from the bathroom, broke my nose, knocked myself out... and I know lots of other folks who have done stuff like this too. What about those kinds of things?' I just looked at him and said, 'Yeah, I know....'"

"I didn't have an answer, but I knew what he was talking about. I mean, I knew that there were lots of accidental injuries like these. So I started asking people questions about that kind of stuff. They were the things that people might tend to call 'stupid' injuries—silly little things—that would usually be no big deal, but instead of nothing happening or nothing much happening, there was an unusually bad consequence. People would tell me, 'I can't believe I broke my wrist' or 'I broke three bones' or 'I broke my ankle'—those kind of things. As it turned out, lots of people (everybody, it seemed) had experienced these 'stupid' little injuries, as they called them. Some of them were also pretty funny. But some of them weren't that little. And some of them that happened in automobiles were hardly anything to be laughing at. But the fact remained, there were a lot of them. It seemed almost everybody had at least one."

"The next month, I was working at an oil field contracting company. I was doing some BBS training, and they were talking about how they'd had a perfect record up until this guy had slipped and fallen getting out of a pickup truck in the parking lot. He'd sprained his wrist and, I think, separated his shoulder or something like that. They were asking, 'How could we have prevented this?' These guys definitely bought into the whole system approach, too. And I remember looking at them saying, 'Well, *you* couldn't have prevented this. But do you think that the worker could have done something himself to prevent it? Think about it: He's gotten out of that pickup truck hundreds of times before without getting hurt. Why did he slip and fall *this time?* That's what we should be asking, not how can we prevent him from slipping and falling out of the pickup. What we should be asking is *why* did he slip and fall this time? What was different this time?'"

"That's pretty much where it started happening for me. I started seeing the futility of thinking that even through observations, you could control things to the level where people didn't slip and fall getting out of trucks, they didn't strain their back by picking something up without thinking about it first, they didn't fall asleep at the wheel, etc., etc."

CHAPTER 5—*Going Beyond Behavior-Based Safety*

Gary nodded, "That's the problem with behavior-based safety (BBS). There was something missing. We were still having the preventable automobile or forklift accidents; we were still having slips, trips and falls, and it's hard to regulate slips, trips and falls. Oh, you can make sure the walking platforms are in really good shape and there's no water or liquid on them, but a lot of these things were just people stumbling over a perfectly clear sidewalk or floor. And we weren't doing anything at all for off-the-job safety. Sure, we were doing a whole lot of talking about doing things for off-the-job safety, but nobody was really doing much for it. And just like the workplace, sprains and strains were an issue there too. You could actually strain your back just reaching over to pick up the newspaper from the front porch."

"At some companies sprains and strains are 40 percent, maybe even 50 percent, of their injuries. Even when you're doing a really good job with all your safety systems, you still have cuts and abrasions—although they may be less serious. In addition, behavior-based safety (BBS) has always been difficult to sustain. It takes a lot of effort, a lot of work, and frankly with the peer-to-peer part, quite often they deviated from looking at behavior to something a little bit simpler, like unsafe conditions. When that happens, you've lost the peer-to-peer observation process, and you may not even know it."

"Not to mention that sustaining that peer-to-peer behavior-based safety process is expensive," I said. "Some behavior-based safety consulting companies recommend taking one hourly employee off the floor to look after and administer the process. So whether you take an employee off the floor or whether that administration comes out of the safety office, you're still looking at a lot of time to administer this. So, yeah, it's expensive. It's also difficult to keep it from going stale, too, especially if you don't have lots of employees, because you tend to be observing the same folks over and over, or they're observing you over and over. It's hard to keep it fresh. But even with a well-run observation and feedback process, there are still problems with slips and falls, sprains and strains, and cuts and contusions."

"Right, those problems were still with us," Gary said. "Lots of people have worked on it, and yet we never had a handle on it. That was the real problem with our concept of 'Utopia' and the behavior-based safety process we were using to try to get there: it failed to address the fact that people make mistakes they never intended to make. We ignored a whole category of issues associated with unintentional error, and without a solution to those issues, you can't get to 'Utopia' no matter how much you want to be there. So are you going to tell us how you actually figured this human error thing out? I mean, what did you do? How did you figure it out anyway? 'Cause lots of safety pros have tried—I know I tried—and every BBS person in the world has tried. So what was it that really turned the corner for you?"

Before I could answer, Gary continued, "There's another question I wanted to ask. Didn't you receive a lot of criticism when you started looking at human error? I mean, for years I was told human error was *never* the problem; it was always the safety system or the hazard. But you were jumping right into human error as being the most significant factor. Didn't you get a lot of negative feedback?"

"I know what you're talking about," I said. "I was fearful of it, too, in that it was almost the equivalent of heresy at that point in time. It was really going against the grain or the popular train of thought."

"When I was sitting there listening to you talk about SafeStart at that advisory board meeting back in 1997, in Dallas," Gary continued, "I realized that you were blazing a trail that nobody had gone down before. When you're at Stage 5, often the only thing left to explain injuries in the workplace is human error. This idea of human error being the primary cause, and then trying to quantify it, or explain it, and then maybe even try to control it, was something that we—the safety profession—hadn't worked on at all."

"Well, like I said, it was definitely going against the grain. The popular way of thinking was that human error was inevitable, so there was no sense in chasing it down. After all, everybody thought the real problem was with the safety system. So I did not think for a second that this was going to be easy. I knew this was going to be an uphill battle all along the way. I knew that there were going to be a lot of safety traditionalists that would be almost waving the flag at me and saying, 'How dare you even hint that human error might actually be a pervasive problem!' I even had one comment from a participant at a national conference say that I was setting safety backwards 30 years."

Gary started laughing. "SafeStart has prevented millions of injuries and thousands of fatalities," he said, "and this guy says you're setting safety back 30 years!"

"I know I was a bit surprised myself."

"Well, we had just spent our careers trying to design systems so that if an error was made, there was a guard there or a barrier or a piece of personal protective equipment," Gary said. "So I guess it was also a bit hypocritical of him to say that. You know, it wasn't like we didn't know human error was part of it. It's just that we didn't know what to do to prevent it. We just ignored it."

I told Gary that I thought he was right. It was almost like everybody had said, "Okay, we all know there's something missing, but we'll agree not to talk about it."

"We'd make the secret handshake and keep moving," Gary agreed.

"Well, yeah, it reminded me a bit of the emperor's new clothes," I said. "We must not let the masses know that there's anything wrong with the system. Ironically, I think that created exactly the type of backlash that safety professionals didn't want. Instead of people buying into the idea that the system was robust enough to prevent all injuries, they had exactly the opposite take on it. As a result, they ended up not trusting either the system or the safety professionals who were telling them the system was enough. I could see why, because if your co-worker was somebody who was always rushing, quite often frustrated, usually tired and very complacent, you might not want to work with them on the bomb squad."

PART 2

Human Factors and Critical Error

Reduction Techniques

CHAPTER 1—*Three Sources of Unexpected Events*

We've seen the development of workplace safety as we've discussed the passage of workers' compensation laws, the rise of organizations dedicated to promoting safety, the beginning of OSHA and the regulations they promulgated, the establishment of credentialed practitioners, the founding of new disciplines, the formalizing of the safety management system, the creation of new safety devices and technologies, and the rise of behavior-based safety processes.

I asked Gary, "Where did all this get us in terms of injury rates and in terms of the way companies were working their way through the five stages?"

"Well, we still have companies in each one of the stages. But those that did go through the five stages are the higher-performing companies. In the United States, the average total recordable rate for private sector employers is around 3.5[1]. The really good

---

[1]The statistics for private industry and manufacturing (in the following paragraph) are taken from Bureau of Labor Statistics, "Workplace Injuries and Illnesses—2010." <http://www.bls.gov/news.release/archives/osh_10202011.pdf>, accessed April 9 2012.

companies, the ones that went all the way through to Stage 5 are below 2.0, and some are below 1.0. So they're doing a really, really good job, but they still struggle with some of those things we mentioned before, like slips, trips and falls, and sprains and strains, those types of incidents."

"But isn't the average total recordable injury rate for manufacturing a little higher?" I asked.

"That's right," Gary said. "I think it's about 4.4 now. The incident rate has actually been in decline for most of the last 15 years. That's probably because of the increased emphasis on safety in most companies. But the companies that have done really well, those that have had a good solid traditional safety program along with an observation-feedback process, we would expect those companies to have total recordables under 2.0."

"I would think that at Stage 5, you might even have companies that were below 1.0. The very top performers would probably be hovering around 0.5 or even below that," I said.

"You're right," Gary replied. "But even the mindset of those companies has changed because they really are trying to become the best. An incident rate of 0.5 that would have been really, really good a few years ago, isn't as spectacular anymore. Now they're trying to even get better. But they understand that every bit of improvement at that level requires a huge amount of effort. They're struggling with figuring out what effort will pay off. That's the thing that can be really frustrating because they've done everything they know of from a technology standpoint. So if they want to get better, every small improvement requires that much more effort. The problem they faced was, 'What and how much do we do?'"

"That reminds me," Gary continued, "I want to talk a little bit about your concepts and techniques, or what eventually became SafeStart. I don't think people want to know that it's in 50+

countries and 30+ languages, and that 2 million people have been trained. That shows the universality of the concepts for sure. But I think what they'd really like to know is the process you went through to figure this out. After all, like I said before, you weren't the only one working on finding the secret to effective injury reduction. In fact, all kinds of people out there were trying: academics, people with grant money, and safety professionals at big companies. They were all trying to figure this thing out. You're pretty much a common, ordinary guy out there working in the trenches. I know you didn't have any research money. So how the heck did you do it?"

The look I gave Gary basically said, "That's not part of the plan." He gave me a look back that said, "I know, but I think it's important." So I thought about it for a second and decided what the heck—maybe he's right, maybe some people would want to know.[2] So...now it was my turn to look out at the lake.

A lot of thoughts flashed quickly through my mind. So many things had happened in the last 25 years starting with how I got into behavior-based safety in the first place. Our company (Electrolab Training Systems) distributed safety training videos. We represented many companies—including DuPont. They had the "STOP" program which, although it had videos, it required some implementation assistance (better known as consulting) because it was much more involved. Because of this, I called the people at DuPont Canada, and they called the people in Wilmington. We decided to meet in Belleville (where Electrolab was) to discuss this issue. Once we sat down to dinner, I quickly excused myself to go to the men's room—but really I went to the bar for a quick smoke (I didn't think any of the DuPont people smoked).

---

[2]If for some reason you're not interested in how SafeStart came about, feel free to skip Part 2 and go immediately to Part 3. It will give you some practical guidelines to help you integrate SafeStart into your safety management system.

When I came back, they were all sort of grinning. I said, "What's up?" Rick Frick (that was his name) who was in charge of training materials in Wilmington said, "While you were gone, we figured it out."

"Great," I said. "So who's going to do it?"

"You are," he said.

"Me?" I almost choked on my salad. (Me? That's funny. I mean, little did these guys know. I wasn't...hadn't exactly been the safest person on the planet.) I looked at my dad (our president). He just nodded, indicating that he'd already said yes.

So, as I used to say to the classes back then, "There's just one more reason to quit smoking...next thing you know, you could end up being a behavior-based safety consultant."

The job got me out west where I got to ski mountains. I also got to do hundreds of training sessions every year, so I got a lot of experience really quickly, which was a plus. However, I was only 28 years old, which wasn't a plus in terms of instant credibility. It wasn't easy talking to people your dad's age about their "behavior," especially when they knew you'd never worked a day of your life in a saw mill or a pulp mill, a refinery or steel mill, or wherever it was. A lot of them would just be looking at me as if to say, "Tell me all about it sunshine." So, like I said, it wasn't easy, sort of trial by fire, but I learned a lot really quickly.

All those thoughts passed in a second or two as I looked at Gary.

"Well, I think the key thing you said is that I was out there on the front lines. Canadian industry being what it is, I wasn't always working in companies who had a lot of their safety management ducks in a row like the oil and gas industry. Over the years, some of those companies had learned lessons the hard way about the integrity of their operation from the very top to the bottom, and conflicting priorities in terms of production over safety. Since a lot of these lessons were world-famous disasters, I'm not giving them any kudos for clairvoyance. But they had more stuff figured out in terms of management responsibility and accountability than in some of the other industries where I was doing training."

"A lot of the companies that I was working with were way down there in Stage 2, and yet they were trying to implement observation processes. When you got on the site, you didn't need to be a CSP to see that there was a lot of room for improvement in terms of engineering controls, machine guarding, lighting, ventilation, personal protective equipment and management 'style.' So, yeah, working in the trenches was pretty accurate. I was doing around 150 to 175 days of training a year, and had 20–30 people in front of me almost every day from 1988 right through to 1999. So if I had a 'hunch,' I could bounce it off one group, then the next group, and then the group after that and so on."

"The first thing in the development of SafeStart that was really significant happened in 1991, and it was totally unintentional. Actually, almost every one of these 'discoveries' were not planned events. I had been involved with selling behavior-based safety since 1984, but I didn't actually start consulting or training until 1988. So now with only three years' experience, I found myself out in the middle of nowhere at a sawmill in Fort St. James, British Columbia. There were about 25 guys in the room, and I was showing them this bit I used to do about the 'Three Sources of Unexpected.' I was relating this to them by saying, 'There are really only three sources: you do something unexpectedly, the other guy does something unexpectedly, or the equipment does something unexpectedly' (see Figure 8)."

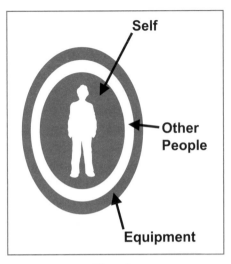

Figure 8
Three Sources of Unexpected

"What I was trying to tell them was that when you were making observations and you saw somebody doing something unsafely like driving a forklift with the load 18 feet (5.5 meters) in the air, that you had a choice: You could say, 'What if you took the corner too quickly and tipped the truck over and hurt yourself badly?' Or you could say, 'What if somebody ran out in front of you suddenly when you had the load up high and you had to hit the brakes really hard?' I was trying to get across the idea that people were only so receptive to suggestions that *they* might make a mistake in the 'self' area, but that human nature being what it is, people would tend to be a lot more receptive to somebody else making a mistake."

"So I was encouraging them to use the 'other guy' making a mistake when they were questioning their co-workers on 'What if something unexpected happened?' If they asked about the 'other guy' there was no illusion of control. In other words, we all know we can't control what the other guy does, whereas we do believe we can control our own error. So—back to Fort St. James—I went through my three sources of unexpected 'bit,' which was one of the few things I had going for me at the time.

The class was getting it—even though I was in the middle of nowhere and it was a rough-and-tumble place that had a horrific injury rate—we were getting along just fine that day. They liked the jokes; they liked the way the course was going."

"It was just one of those days where I felt comfortable enough to try something new. I think every trainer who's done a few hundred days of training experiences this, where every now and again you're feeling good, so you try something a little bit new. If it's good you keep it, and if it isn't you don't. Well, this was one of those days. I just looked at the class and said, 'Let me see if I can give you some additional perspective on this.' I pointed to the Three Sources of Unexpected diagram that the overhead projector was showing on the screen. 'There's about 25 of us here now, so how many of you guys have been seriously hurt—we'll cut it off, say, at stitches or worse, just to make sure you can remember it—how many of you guys here have been seriously hurt because the equipment or the car you were driving broke, malfunctioned or did something screwy, unexpectedly—like the wire rope snapped, the hose burst, the seal failed, the traffic light started working incorrectly—that kind of stuff?'"

"*Nobody* in the room put their hand up. And then I said, 'Well, guys, from what I've been told, sudden equipment failure or component malfunction is usually the primary cause in about one to five percent of all incidents. There are only 25 of us here, but if we had 200 or 300 people in the room, I'm pretty sure we'd see a few hands in the air.' Then I pointed at the 'other guy' area and said, 'Now if we exclude contact sports, bar fights, things like that, how many of you have been seriously hurt because the 'other guy' did something unexpectedly? You know, skidded through the intersection and T-boned your car, that kind of thing.' Once again, *nobody* in the room put their hand up. Now we're all sort of looking at each other, and it was really quiet. Then one of the guys in the middle of the room looked behind him to make sure nobody had their hand up. He turned to the front and said, 'So...' he paused and turned to look around the room again...'We all hurt ourselves!' All of us were sort of stunned. I don't think any of us expected to end up where we did."

Then Gary intervened. "So you expected all the hands to go up for the 'other guy.' Was that the point you were trying to make?"

"Yeah, that's exactly what I thought would happen. I expected everybody would put their hand up because they had been hurt by the other guy doing something unexpectedly. Then I'd be able to say, 'So, as you can see, there's a significant amount of risk with the other guy that you can't control, which is one of the reasons why you need to use the *other guy* strategy when you're observing and talking to people, because it's going to make your observations go so much easier.' But you're right, Gary. I never got to do any of that at all! And now, I'm just floored at this because I knew that if you asked 25 randomly selected people a question and everyone answered it the same way, you could be pretty sure that most everyone else would also give the same answer."[3]

"Now, you probably wouldn't say that this group of sawmill workers up in the middle of B.C. was totally representative of cosmopolitan society in North America, and you might wonder if they were a truly random group. Nevertheless, it was a significant event, and I knew it, and everybody else in the room knew it too. I remember this guy standing up in the back; I think his name was Kenny; he was the plant engineer; he wasn't a big guy. He just said, 'Larry...is this the way it is everywhere?' And I just looked at these guys and said, 'I don't know.' Then I think I said, 'Actually, I barely know what I'm doing, if you want to know the truth. They gave me some overhead transparencies and a bit of training, and I grabbed my skis and boots and started getting on planes. Heck guys, I've got a degree in economics, and I spent

---

[3]Creative Research System, "Sample Size Calculator," < http://www.surveysystem.com /sscalc.htm>, accessed April 9, 2012. Note the paragraph entitled "Percentage": "If 99% of your sample said 'Yes' and 1% said 'No,' the chances of error are remote, irrespective of sample size."

the last five years in California modeling, doing soap operas and TV commercials.' Then I quickly continued, 'But I get to do hundreds of these sessions a year. Maybe I should start trying to figure this thing out.'"

"So from there on, in every class I taught, I started asking more and more questions. I would show them the Three Sources of Unexpected diagram and talk about the 'self' area. It was a real eye opener for people everywhere I went—that it wasn't the equipment, that it wasn't the other guy. Even when you included all the stuff on the highway, it hardly ever was the other guy. Knowing that I had this bit that I could do, that there would be a 'eureka' effect with every class I taught, really helped build my self-confidence because I didn't have all that much experience. It was like a 'feather in my cap.' It also improved the quality of the classes. I'd open and establish the 'self' area as being well over 95 percent of all the injuries that have happened to all of us in the room. Then I would start asking people questions about how they had been hurt. I also learned a lot going out with the guys for smoke breaks, where you could ask casual questions, where nobody really worried that the answers they gave you might get back to their boss."

"So that's how it all began?" Gary asked.

"As unlikely as it sounds, yes, that was the beginning," I replied, "and it led to a whole lot of other questions."

CHAPTER 2—*High Risk Versus Real Injuries*

"The one question that I got a ton of mileage out of was, 'What's the most dangerous thing you've ever done?' And let me tell you, the answers to that rarely—if ever—were things they did at work. In a huge percentage of these stories, as you can imagine, copious quantities of alcohol were involved. (That was probably good because given just how stupid some of these things were, you would almost want to be drunk, otherwise you'd just have no excuse.) However, apart from the 'so drunk I was seeing double' stories, I thought the answers I got were interesting, because the way most people talked about workplace safety back then was that by far and away the most dangerous place on the planet was where you worked. I mean, this was where you *really* needed to be afraid. Yet when you started asking people about the most dangerous thing they'd ever done, you hardly ever got a workplace story."

"When I started asking people about high-risk activities, they would start talking about riding snowmobiles down the hill at 120 miles an hour (193 km/h), driving their cars over 150 miles per hour (241 km/h), kayaking down whitewater rapids, skydiving, scuba diving with sharks...those kinds of things. Then I would ask them—this is where the thing really became interesting—'What's the worst injury you've ever had, not including contact sports?' What I'd find is that in most cases, their worst injuries didn't come from the most dangerous things that they'd ever done. Rather, their worst injuries came from—and these are actual quotes here—'It was just one of those silly things,' or 'I slipped and fell, but I hit my head,' or 'I hit my elbow and it shattered.' Quite often, their worst injuries were just unlikely or unlucky outcomes from very common events."

"I'd also hear things like, 'I slipped and fell down, but the result of the fall was actually quite serious.' Then they would hasten to say they broke a bone in their ankle, or they blew their knee out, or they hit their head, knocked themselves out and woke up in the hospital unable to move their legs. But they weren't doing anything where they were consciously aware of the risk or consciously aware of the fact that they were deliberately increasing the risk. For example, they weren't going 100 miles an hour (161 km/h), but rather they fell asleep at 60 miles an hour (97 km/h) and nearly killed themselves. This was really interesting to me because I had also done many (too many) dangerous things myself, and yet one of the worst times I'd ever been hurt was when I fell asleep at the wheel. On that occasion, I wasn't speeding. I was only going about 60 miles an hour (97 km/h). Not only was I surprised that the most dangerous things didn't equal the worst injuries—which was kind of the 'party line' at the time—but I was also surprised that what had happened to me had also happened to a lot of other people. So I kept asking questions to people at the classes I taught, questions about how they had been hurt, either on or off the job."

"Were the conversations with all these people all over the map, or was there any consistent thread that went through them?" Gary asked.

"No, at this point in time there was no pattern, no thread other than what I already said, which was that the most dangerous things they did weren't what caused their most serious injuries. It was just a lot of information, none of it really fit into any particular pattern. But I was hearing lots and lots of stories."

"By this stage of the game, I firmly believed that there was a missing piece to the safety management puzzle, and that the standard party line about the traditional, compliance-based safety management system being all that was needed to prevent injuries was not completely legitimate. I *knew* that there was something missing. It felt a bit like *The Emperor's New Clothes*, and I *knew* that there had to be other people who knew there was something missing too. But not everybody was open to talking about it."

"So the first breakthrough involved the 'self' area and the second was the 'most dangerous thing versus the worst injury,'" Gary nodded. "Okay, what came next?"

CHAPTER 3—*The First Human Factor: Rushing*

"I guess the next breakthrough came when I went to a conference in Las Vegas. Now, I went there mostly because I wanted to go to Las Vegas. But by then, I also wanted to do more than just go to the exhibit hall and talk to suppliers of videos, which was really all I was supposed to do. By this time, I had also started attending the conference's technical sessions. As I looked through the program, I saw a session on 'Using Quality Improvement Methods to Improve Your Safety Performance.' The session piqued my interest because I had done a lot of quality control training (Deming, Crosby, Juran, Taguchi) in the mid-'80s for the automotive industry. That's actually how I became the candidate for the behavior-based safety position. People reasoned like this: 'Well safety and quality are a lot alike. Since Larry has been doing the quality training, why don't we get him to do behavior-based safety training in Canada?' Of course, I had no idea at the time how people would react. You can talk all day long about quality, and you might put people to sleep, but nobody gets mad at you. Whereas you can barely open your mouth to talk about behavior-based safety, and you have people take an almost instant dislike to the message (and the messenger). It's really quite interesting."

"Anyhow, I attended this session just to hear what the presenter had to say. The session featured a guy named Brooks Carder who talked about a really interesting case study. His consulting firm got a job with a large pharmaceutical company to interview a cross section of their people from all areas of the operation and all different management levels. After ensuring the data was anonymous, they would present the findings to senior management with sort of a 'read it and weep...this is what's really going on at your place' type of presentation. The management group at this company knew that they were having quality problems, but they didn't know why. So they hired Carder's firm to come in and interview their employees."

"What Carder found was that when the plant got rush orders, instead of the plant following good manufacturing practices (GMP) for the batch they were making—I'm not sure just what it was—they simply added all of the ingredients in at the same time, mixed them up and shipped the product out. They thought everything was okay, everybody was cool, and they weren't aware it was causing any problems. But eventually, the fact that there were quality problems made its way back to the plant. Carder told us that what the management group was thinking of doing—their obvious solution—was to round up all the employees and tell them something like, 'Either you follow procedures or you are out of here!' You know, just read them the riot act, no deviation, no way, no how, the usual standard deal. But what Brooks Carder advised them to do was to look at everything, look at the whole system. He was practicing what Deming taught: Employees account for 15 percent of variation at most, whereas the system is usually responsible for 85 percent of the quality problems."[4]

"When they looked at the whole system, they found that the reason the plant was getting rush orders was because of a slow paperwork system. Orders were taking three or four weeks, sometimes months, to hit the plant floor. So they set about to

[4]Roger W. Berger et. al., *The Certified Quality Engineer*, 2d ed., American Society for Quality, Quality Press, Milwaukee, 2007, p. 4.

improve the paperwork process. I'm not sure exactly what they did, but I think this was back in the very early days of cell phones. When they placed their orders by phone, the orders made it to the plant floor in a couple of days, sometimes even in a couple of hours. As a result, the plant didn't have to rush, and they reduced their quality problems by more than 50 percent, In addition, Brooks said that they also reduced their injuries by 50 percent. He concluded the session by saying something like, 'Don't just look at the employee, look at the whole system, and you may find huge improvements in your system that would provide a much higher yield than just simply looking at the behavior of the employees.' Everyone clapped, including me."

"It wasn't until I was flying home the next day that it hit me that there were about 200 people at this session. It was one of the sessions at the American Society of Safety Engineers' conference, and not *one* person there asked any qualifying questions like, 'Are they lost-time injuries? Are they medical aids? Are they recordable injuries? Are they first aid injuries, or are they all of the injuries put together?' Nobody asked a single qualifying question. You know the ASSE people. This is not the kind of group where you can just float a 50 percent reduction in injuries, without any data or anything to back it up, and not have anybody ask you a qualifying question, especially if they suspected something was amiss."

"Well, I am part of that group," Gary said. "But now that you've told the story, I can see it happening because it just intuitively makes sense to me that rushing would be an issue with respect to safety. So I guess it also made sense to them, too. That's probably why nobody asked any qualifying questions."

"Right," I said. "We all just accepted—at face value—that rushing contributed to 50 percent of the accidental injuries. So what I started doing, the very next week with the first class I had, was to ask, 'How many of you here have been hurt, at least once in your life, on or off the job, because you've been in a rush?' Just as you would imagine, everybody in the room put their hand up."

"Interesting," Gary said.

"Actually," I said, "I had lots of people putting both hands up and lots of laughter. People were freely admitting that rushing had been a *big* problem. So, now I started thinking, okay, rushing is huge. But rushing alone wasn't really enough to get you hurt. I knew about 'eyes on path' from behavior-based safety and putting critical behavior checklists together for clients. I also knew about line-of-fire from the work I had done with behavior-based safety. So I started thinking about eyes on path, and I realized it wasn't just path, it wasn't just walking—it was eyes on *task*. So I had rushing, eyes on task, and then line-of-fire. I knew about fatigue from personal experience. I'd totaled a company car in 1986 when I fell asleep at the wheel and nearly killed myself. So I now had rushing, fatigue, eyes not on task and line-of-fire."

"Okay, so you've got these four things figured out: rushing, fatigue, eyes on task and line-of-fire. What did you do with them? I mean, that's a good start, but what did you do with those concepts?" Gary asked.

"Well, not much initially, other than to ask lots of questions about those four things. After I had defined the 'self' area, I would ask, 'Can you think of a time you've been hurt in the *self* area where you weren't rushing, you weren't tired, you had your eyes on task, and you were aware or thinking about the line-of-fire or what could be coming at you?' What I was finding was that you didn't have many hands in the air. You'd see the odd hand going up, but you'd also see a lot of people who were starting to realize that these four things were involved in a huge percentage of the times that they had been hurt."

"Okay, so, these applied to the 'self' area," Gary said. "But they may not apply to the equipment area or the other guy."

"Right," I replied. "So you'd have to start off going through the three sources of unexpected and defining the 'self' area for people first, because you would have to get them to realize how few examples there were—not just for themselves, which usually would be zero—but how few there were for other people, too. You see, we look at sensational reports on TV and in the newspapers of events that aren't caused by human error, like tidal waves and earthquakes, and our perspective about how most people really do get hurt is skewed. So you have to start off by explaining the three sources of unexpected events and get them to see how significant the 'self' area is."

CHAPTER 4—*Personal Risk Pyramids: Why Everyone Thinks They're Safe Enough*

"So where did the 'three sources of unexpected' concept originally come from?" Gary asked.

"It was around 1988 or 1989," I said, "but I'm not exactly sure where it occurred to me that there were only so many ways that unexpected things could happen: *You* did something unexpectedly, *somebody else* did something unexpectedly, or the *equipment* did something unexpectedly. When I taught it in class, I'd tell everybody that we weren't talking about tornadoes, volcanoes, earthquakes, floods, tidal waves or hurricanes. After all, there's not a whole lot you can do about them. Everybody got the idea; it made sense to them. Back then, I wasn't thinking about the three sources of unexpected in terms of percentages. What I was using the three sources of unexpected for was just to make it easier to talk to someone about deliberate risk. That was where the 'other guy' became really useful, because from a human nature point of view, people are more receptive to the other guy making mistakes than themselves."

"However, by 1991, I started getting people to build their own personal injury risk pyramids. I originally started just asking people about how they had been hurt—'What's the most dangerous thing you've ever done?' or 'What's the worst injury you've ever had?' But now, I'm actively trying to quantify things by asking, 'How many major injuries have you had over the course of your lifetime on or off the job? How many minor injuries—things like stitches, sprains and strains?' Then I'd ask, 'How many minimal injuries like cuts, bruises, bumps and scrapes?' When I asked about those, I'd hear the whole room groan and say 'Hundreds' or 'Thousands.'"

"When we started having kids, it became obvious to me this was not a static model. In other words, there was a real dynamic to all of this: You got hurt many more times at the beginning of your life than you did towards the middle and end of your life. I started noticing that children would have many more bruises on their shins than adults. As it turns out, little kids get hurt about 52 times more frequently than adults. No wonder most adults think they're safe enough. We've all improved about 5,000 percent from when we were little.[5] I started trying to keep track of the injury data attendees were willing to share. So now I'm not just asking people about how they've been hurt, I'm also trying to keep track of just *how often* people have gotten hurt and *how serious* the injuries were."

"I was also starting to ask people about significant close calls: 'How many times do you think, if circumstances were slightly different or if your luck had run out, you could have died? This would include: every time you have fallen asleep at the wheel even if it was just one of those really long blinks, every time you fell backwards off of your bicycle without wearing a helmet, every time you fell backwards where something hard or sharp was behind you, and all the crazy stuff you did in high school and college with your friends.' The results, which are based on

---

[5]Little children get hurt about 20 times per week. Adults get hurt about 20 times per year. From 20 a week to 20 a year is a 5,000 percent improvement.

the responses of 150,000 adults in Canada, the U.S., Mexico, Europe, South America, Australia, Asia and the Middle East, are shown in Figure 9."

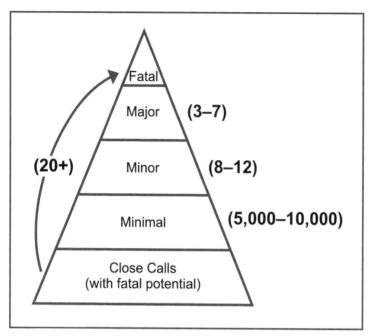

Figure 9
Average Injuries and
Potentially Fatal Close Calls

"I was still trying to make some sense of all the data I was collecting. But at this time, I still only had rushing, fatigue, eyes not on task and line-of-fire, and I knew that the 'self' area was well over 95 percent."

"Okay, those numbers are very interesting, but I know that there's more involved. When did you add other states and errors? How did that come about?" Gary asked.

"Like I said before, I got a lot of help from the people in the training sessions I conducted. At the end of the class I'd just ask, 'Can you think of a time you've been hurt in the 'self' area when you weren't rushing, you weren't tired, you were watching what you were doing and you were aware of the line-of-fire?' What I got a lot of, very quickly, was, 'I just slipped and fell,' or 'It just dropped out of my hand,' or 'I just started skidding.' It didn't take long before I realized that there were also problems with balance, traction or grip. So now I've got five things: rushing, fatigue, eyes not on task, line-of-fire, and balance, traction or grip."

CHAPTER 5—*Four States and Four Errors*

"Okay, it's starting to come together," Gary said. "You've got five factors. So that's when...about 1995?"

"Yeah, about 1995 or 1996," I said. "I had five: two states and three critical errors. I had them printed on small cards, and I'd pass the cards out at the end of the two-day class. Then I'd ask the question, 'In the *self* area, can you think of a time you've been hurt when these things were not a factor...?"

Gary interrupted. "So now you've got five factors. It looks like you've got the beginning of the pattern here. What did you do with the five?"

"I really didn't have the pattern, though. That was just it," I said. "I had these five things, but they weren't in a logical order, although I didn't know it at the time. I had rushing, and then I had eyes not on task and line-of-fire. Next I had fatigue, and then I had balance, traction and grip (see Figure 10). So I had these five things which I just put in the order I had discovered them. I was testing the five all the time. I'd ask, 'Can you tell me a time you've been hurt in the 'self' area where one of these five things wasn't a factor?' Most of the people kept their hand down because they didn't have an example that fell outside of those five factors. But during many classes, it seemed like I'd get one or two who would raise their hand to give me an odd exception."

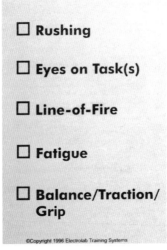

Figure 10
Early SafeStart Card

"When I'd ask these people what happened, they would tell me stories about when they weren't paying attention to what they were doing. But after hearing enough stories, I'd ask, 'Well, why weren't you paying attention to what you were doing?' In listening to their answers, it dawned on me that rushing or fatigue weren't always involved. If you were fearful of the hazards of a task, if you were still scared, you would automatically be paying attention, in fact, you couldn't help but pay attention to what you were doing. But as soon as you'd done a task enough times that the hazards didn't actually make you fearful anymore, your mind could wander a bit. So from these exceptions, I actually discovered complacency and mind not on task."

"Okay, now you've got seven factors. How did you get the last state?" Gary asked.

"Well, there's quite a story of how I discovered the frustration thing in the fall of 1996. Years before in 1991 and 1992, I had done a lot of work for a utility in western Alberta. They'd really liked the overview sessions that I had taught and done very well with the observation process. In spite of the fact that I was just starting out, they gave me hundreds of days of work. That work gave me an opportunity to develop both my public speaking and personal communication skills. Being a behavior-based safety consultant is more than just knowing facts, it has a lot to do with the way you deliver them. Talking to people about their behavior as it relates to safety requires a bit of humor, and you have to be able to be somewhat entertaining."

"So with all the work they gave me, it was almost like they footed the bill for my apprenticeship; there was a long, positive relationship. I was very thankful for them, and they were very thankful for the two-plus years the three generating stations went without a lost-time accident. They'd had seven to nine lost-time accidents per generating station per year before, and now they were running all three with zero."

"Four years later, I returned to one of the generating stations. I knew that they were having some trouble with safety. But I thought that we still had the rosy relationship we had two or three years before. I knew that the company was basically trying to just keep this generating plant running for a while longer until it could be mothballed and taken out of service. That's because there was a lot of asbestos in the plant, and it would have cost more than it was worth to remove it. They were basically not trying to let the place fall apart, but they were certainly not spending a lot of money on upkeep. So, once again, they want me to do some behavior-based safety training. But, you can probably see what the workforce was thinking, 'Hey, there are a lot of unsafe conditions here.' They're also worried about their jobs, right? 'What am I going to do for work when they close this place down?' The company hadn't told them exactly when the facility would shut down."

"I arrived knowing that there might be some issues. They told me the training room was downstairs in the basement. I'd never been to this place before because the training sessions I had done three years ago were all at the main generating station a few miles away. When I went down to the basement, I saw a cartoon on the training room door (one of the employees who worked there must have been an artist). The caption read, "Behavior-Based Safety Training by Electrolab Training Systems." It wasn't the *actual* name of the program we implemented, but it was the name of the company I worked for. The cartoon showed a guy with electrodes coming out of his head and somebody else zapping him with electricity. It was funny. I mean, I had to laugh. But on the other hand, it was a bit of a cheap shot, especially considering how much good we—my company and I—had done for them. But it's all right, it's funny enough."

"When I walked into the training room, I saw a few people sitting in the room already, people who start getting paid at seven o'clock even though the class didn't start until 8:00. So when I got there to set up about 7:30, these guys from the maintenance department were already there. I got out all my materials, put the first transparency on the overhead projector and turned on the power. But the overhead projector didn't work. So I plugged something else into the electrical outlet to make sure that there was power there. There was."

"I didn't know what the problem could be, other than maybe the bulb was burned out. But there was a spare bulb inside the projector, which was pretty standard feature for overhead projectors back then. So I put the new bulb in, plugged the cord back in and turned it on, but it still didn't work. And now the only thing I knew how to do was the old TV thing—start banging on it. So I started banging on the overhead projector. As my blood pressure continued to rise, I noticed one of the maintenance guys watching me and just kind of smirking. Between the cartoon and the overhead projector not working, right then I was kind of mad. I like things to go well at the beginning of a session. Now everyone was filing into the classroom, and I didn't know how to get a new overhead projector. As I continued banging on the overhead, the guy who was smirking at me before came up to me and said in a very wise but condescending tone of voice, 'You can't hurt the equipment, but the equipment can hurt you.' Then he unplugged the projector, took it apart, showed me a wire that was disconnected, fixed it, put it back together, plugged it back in and it worked. Now, I never did find out whether he was the one who disconnected the wire in the first place, but it was one of the quickest fixes I'd ever witnessed in my life."

"Sure makes you wonder whether it was a coincidence or not," Gary said. It was more a statement than a question.

"Well, I don't think it was a coincidence. Like I said, I don't know whether he was the one who did it or not. But when the class started, I just asked them, 'How many of you people here have been hurt because you've been frustrated with something that wouldn't work?' Everybody in maintenance put their hand up. Then I said, 'How about with people? How many of you have been hurt because you've been frustrated with somebody?' Everyone raised their hand. Finally, I asked, 'How many of you have been so mad at somebody or something that you have either punched or kicked an inanimate object hard enough to break a bone?' There were only twenty of us in the room, but there were four hands in the air, and mine was one of them. I should have been able to figure out that frustration was a contributing factor just from what had happened to me in the past, but I hadn't thought about it until that day with the overhead projector."

Looking at a finger on his left hand, Gary said, "Let me tell you about an experience I had long before I started talking to my people about rushing and line-of-fire. I was working for a major manufacturer many years ago; I was a rookie, and I was assembling some equipment. It was just a short-term job, and I had no intention of staying there, I worked for this supervisor by the name of Frank. I didn't care for Frank; I didn't care for the job; I was there just for the money."

"At the plant we had this interesting rule: If the next shift person on the line wasn't coming in for some reason—they'd called in sick or something—the first shift person had to stay over two hours to give the supervisor an opportunity to replace the person who had called in. It was a big issue back then; we're talking about ancient history back in the 1960s."

"One day right in between the end of one shift and the start of another, I think it was about 3:20 in the afternoon, I had taken my safety glasses off and was sitting down on a folding chair—basically I'm getting ready to go. Frank came up and he said, 'Hey, Hig, you gotta stay over a couple of hours tonight,' and I said, 'Well, Frank, I really don't have time, I've got something I really need to do.' But Frank was really insistent on it. So I told him, 'Frank, I'm on the beginning point of the assembly line where everybody who works on the line starts. There are dozens of other men here who have plenty of experience doing my job. Besides Christmas time is getting close, and most of these men are married and they have children. I'm sure they would love to work a couple hours' overtime. Why don't you ask them?' Frank responded in a very 'not nice' way, 'No, I'm not asking them. I'm not…I don't have to ask them. I'm telling *you* that you have to work!'"

"So an argument ensued; I'm yelling at him—well, maybe not yelling, but certainly we were raising our voices. He's telling me I've got to work. I'm telling him I can't work, and I don't want to work, and it doesn't make sense that he's making me work. He's telling me it's the rule. Finally he stopped, and he asked, 'Well, what is so important about you getting out of here today?' And I said, 'Well, Frank, if you *must* know, I have a date'—that's why this story is ancient history. Frank said, 'Well, you're going to get out of here by 5:30, so you've got plenty of time to go on a date.' I said, 'Well Frank, you *don't* understand. I wanted to go out with this girl in high school, but she didn't want anything to do with me back then. Now she's decided maybe she will go out with me. However, she's moved out of town—at least an hour and a half away. So,' I said, 'by the time I get out of work at 5:30, take a shower and drive all the way up there to take her out for dinner, it's not going to work.'"

"He said—I'll never forget the words—he said, 'I don't care if you've got a date with Ann-Margret. You're not going. If you don't work these two hours, don't worry about coming into work tomorrow 'cause you don't have a job!' Well, if I *had* a date with Ann-Margret, I would have quit on the spot, but I didn't have a date with Ann-Margret. The argument kept going back and forth, and I felt I was losing. So I got up off the folding chair, stood up and in a fit I yelled, 'All right, Frank, I'll work!' and I slammed the folding chair shut. Unfortunately, when I slammed the folding chair shut, I caught the middle finger of my left hand in the hinge and literally cut off about a half-inch (1.3 centimeters) of the end of my finger. After it happened we both just stood there. Frank was looking at part of my finger on the floor. I was looking at part of my finger on the floor. Eventually, we picked it up, and obviously, I headed to the hospital. Here, take a look. See the nice job they did stitching all the way around it."

I looked at his finger and nodded. "You're right. They did do a good job of stitching it back on there, so you can still signal other drivers and...."

"That's right," Gary quickly interrupted. "However, I can't feel a whole lot with it even today, more than 40 years later. But I guess I showed Frank—I didn't work."

I told Gary, "This is exactly what would happen with so many people when I would get these eight factors out there—whether it was frustration, or the phrase, 'eyes not on task,' or 'line-of-fire,' or 'mind not on task,' or maybe even complacency—it would trigger a story. They would remember something, and it would be like 'Okay, I can see one of my injuries in what you just said right there.' In other words, yes, Frank was making you mad, but he didn't do anything unexpectedly, and the folding chair didn't do anything unexpectedly."

"No, but I certainly didn't recognize that I was in the line-of-fire, either."

"So now I've got all eight factors," I continued. "But they're not in any order, and I'm testing all eight. Now when I asked people about rushing, eyes not on task, line-of-fire, fatigue, balance, traction, grip, complacency, mind not on task and frustration, I'm not getting *any* hands up in the air. And I'm asking the question in class after class after class. As long as I defined the 'self' area, when I list the eight factors up on the flipchart, and ask about their experience, there are *no* exceptions to these factors."

CHAPTER 6—*The Risk Pattern*

"Okay," Gary said. "You've got these eight factors, and you're talking about the 'self' area. So did you start testing those eight things out more? I mean, did anything change? What did you do with the eight? You've got a big sample size now; you've got a lot of data. Granted it wasn't a formal research project, but when you've got a lot of people not raising their hands, you know you're close. So what did you do with that information?"

"Well, that's exactly it," I said. "I'm going through class after class trying to figure out what I should be doing with this. I mean, these eight factors are almost a hundred percent of the injuries in the 'self' area, which was over 95 percent of all injuries—on or off the job. I'd already asked well over 4,000 people about the eight factors because I was starting to speak at safety conferences. But I didn't know what to do with them. So to answer your question, 'What did you do with it?' Well, that's just it. I had it and I'm thinking, 'This is cool, except that I don't really know what to do with it.'"

"By now it's 1996, and I had just finished doing a behavior-based safety session for a major oil company and one of their sub-contractors up in northeastern Alberta. The session had gone very well. A few days later I called the office to speak with the account manager who looked after that industrial sector; she had been at the session. We talked on the phone about how the session had gone. She told me the company really liked it, and they were booking more sessions. She wanted me to check my calendar and pencil in a few dates. Finally, I asked her, 'What did you think when I presented those eight factors in the 'self' area and nobody could put their hand up?' She said, 'Well, it was really interesting for me because even though you've told me about it before, it was the first time I've seen it. But,' she said, 'I thought it was also really interesting for them because I thought it made them realize just how often it wasn't the equipment.' 'No, I didn't mean that,' I replied, 'I know that already. But what about when they couldn't find a time when it wasn't one of these eight factors?' And she said, 'Yeah, that was okay, I guess, but I don't know what it means. I hear the eight things, and I can't think of a time myself, but I don't know what you're going to do with it.' 'Yeah,' I paused, 'I don't know what to do with it either.'"

"While I was talking with her on the phone, I was sort of doodling on a notepad. I had the eight things written down, and then I started rewriting them. For whatever reason, I wrote the four states in one column, and then I wrote the four errors in another column. And while I was telling her, 'I don't know what to do about this, either,' I just drew an arrow between the four states and the four errors, and the whole thing just sort of happened for me right then."

"I said, 'Lara, I've got it! One or more of these four states causes or contributes to one or more of these four errors. Rushing by itself doesn't get you hurt; it just contributes to you making one of these errors.' She paused and said, 'Now *that* makes sense.' And I said, 'You could learn to recognize these four states so you didn't make the error...' So that was how I got—or stumbled upon—the pattern (see Figure 11)."

"Okay," Gary said, "so the state would lead to the error. But then when you make the error it doesn't mean that you're going to get hurt, does it?"

Figure 11
State-to-Error Risk Pattern

"No," I replied. "Just because you make an error doesn't mean you'll get hurt. It just means you've increased the risk of getting hurt, especially if you're moving." Trying to finish the story, I continued, "After I thought I understood the relationship between states and errors, I needed to test it. So now what I did at the classes in the next few weeks was that I would define the 'self' area, and then I would just ask about the four errors. 'Can you think of a time you've been hurt in the *self* area where you had your eye on the ball, your mind on the game, you were aware of the line-of-fire, and you were at least conscious of losing your balance, traction, or grip?' *Nobody* put their hand up. And then I'd say, 'Okay, now, let me ask you if you can think of a time in the *self* area where you've been hurt when you weren't rushing, you weren't frustrated, you weren't overly tired, and you hadn't become so complacent with the hazards that you just weren't thinking about them at that instant?' And again, *nobody* put their hand up."

"What I was quickly establishing was that those four states caused or contributed to those four errors. And like you said, once you made the error it didn't mean you got hurt, it just meant you increased the risk of getting hurt. So now I've got a state-to-error risk pattern, and I'm thinking that once you make the error it's too late, right? Then it's just a matter of luck and how much hazardous energy you're dealing with. In other words, a 3-foot (1-meter) fall is better than a 30-foot (9-meter) fall, and landing on your feet is better than landing on your head. But by and large, once you make one of those four critical errors or a combination of them, now it's just a matter of luck and the overall amount of hazardous energy you're dealing with. Making mistakes with household current at 120 volts is one thing; making mistakes with 13,800 volts is another. But in both cases, you weren't trying to make any mistakes in the first place. So if you could get people to recognize the state and then think about not making a critical error, you'd be much less likely to make one."

CHAPTER 7—*The First Critical Error Reduction Technique (CERT)*

"That must be about the same time you and I ran into each other at that safety advisory board meeting in Dallas," Gary said, "because I remember you talking about states and errors, and I think you used the word 'triggering.' I believe that occurred around April of 1997. So did you actually have everything figured out by then?"

"Not really," I said. "By that time I'd worked out the four states and the four errors and understood the state-to-error risk pattern pretty well. As you just said, I had figured out the first of what would later be called critical error reduction techniques or CERTs. It's almost a no-brainer once you understand the state-to-error risk pattern. You see, in most cases a person is in one of the states before they make an error. That means you could use the state as a trigger. For example, as soon as you recognized you were in a rush, you should try to slow down. But if that's not possible, instead of thinking about why you were rushing or what was going to happen to you if you were late, which is what most people do, the idea would be to get them to quickly start thinking about eyes not on task, mind not on task, line-of-fire, and balance, traction, grip. So that's the first CERT: Self-trigger on the state (or the amount of hazardous energy) so you don't make a critical error. Usually, just thinking about not making a critical error is enough to keep from making one. But this first technique only works with rushing, frustration and fatigue."

"You also asked us your famous question at that meeting: 'Can you think of a time you've been hurt when one of those four states did not cause or contribute to one of those four errors in the *self* area?'" Gary said, "I remember you holding up a $20 bill when you asked the question. How did that get started anyway?"

"Once I had the state-to-error risk pattern, I went and tried my very best to test it. I started off just asking people. Like I said before, there were no hands in the air for either the four states or the four errors, provided I'd defined the 'self' area properly. But I knew that this 'no hands in the air' thing wasn't very conclusive. As you said earlier, I didn't have access to research grants or anything like that. Plus, I didn't know how much effort people in the class were putting into it. I would hopefully have established some rapport with them during the day, and usually I'd ask the questions about the four states and the four critical errors at the end of the day, sometimes at the end of day two. So I started worrying that perhaps they weren't thinking that hard about a time they had been hurt when one of these four states had not led to one of these four errors. Maybe they were just trying to humor me or they just wanted to get out of the room. So, one day I was asking a class about the four states and the four errors at the end of the day. It was about five minutes to four when I asked them the question. As I passed out the little cards, everybody almost instantly stood up and started walking out the door. I don't know why I did this—probably a bit of frustration—but when about half of them were out the door, I said, 'Tell you what. I'll give any one of you $10 right now if you can give me an example, not including sports injuries —without making it up—where one of these four states didn't cause one of these four errors."

"Did you have any takers?" Gary asked. "Ten bucks is ten bucks is ten bucks."

"No, but they all turned around and came back into the room. Then they looked at the card again, and looked at each other, and then they looked at the card again. And then there was sort of a collective 'Holy smoke!' from the whole group—including me. And I said, 'Okay, guys, thanks a lot, really appreciate it.' I repeated the offer in the next class, and the next class, and the next class.... And after I'd asked about 1,000 people, I upped it to $20. After a few more months, I'd probably asked 1,000 more people."

"So by the time I had met you, I had already tested the four states leading to the four critical errors with $20 in my hand in front of about 5,000 folks. When you saw me do this at the safety advisory board meeting, I was pretty confident that no one was going to be able to get that $20 from me."

"Yeah, I remember when you did that. We were all thinking pretty hard."

"Now, if you remember, Mark, the guy who used to be a volunteer fireman, did have an example about going into the warehouse where, unknown to the firefighters, magnesium was stored. When water hit the magnesium, it blew him 100 feet (30 meters) out of the building into the fire truck and broke his leg. I do recall looking at everyone else in the room and asking if anybody else had done the human cannonball thing 100 feet into a fire truck. We all just started laughing. I actually ended up using that example as one of only four examples that I got out of about 20,000 people that had an exception to that state-to-error pattern."

"Yeah, I remember that meeting well," Gary said. "But it was that state-to-error relationship that got me thinking about your concepts and techniques. And the next week I invited you to come up to Iowa and do some...I guess I would call them, employee safety motivation sessions."

CHAPTER 8—*Impact on Workplace Injuries and Off-the-Job Injuries*

As I listened to Gary, I smiled and said, "What I recall is slightly different than that. I remember you asking me if I was going to the conference in New Orleans, and I said yes. I think it was the ASSE. Then you said, 'Any chance I could get you to stop in Iowa on your way back and talk to a few of our folks about these concepts?' I remember saying, 'Sure, if you pay the difference in the plane fare and pick up the hotel, car rental and meals.' You said, 'No problem.' Then a couple of days later you phoned me again and you asked, 'How much do you charge a day? We want to pay you.'"

"That should have been my first clue right there that this wasn't going to be just a few of your folks. So I said, '$1,500 a day'—the going rate back then—'plus expenses.' You said, 'No problem,' which was a good thing because $1,500 was really kind of top dollar at the time. I was surprised there was no real resistance to the cost. That should have been another clue that perhaps this wasn't going to be a cakewalk. The next thing I remember you told me is that you were planning on having 400 people per session and four sessions a day in order to get through 2,800 people in two days."

"I thought it was a great idea, and I want to remind you that I paid for it," said Gary. "The people at the plant had already been hearing a little bit about rushing and eyes on task from me, but you had it all put together. My safety staff thought it was a great idea too, so we got it approved by the plant manager. It wasn't cheap for us to pull 400 people at a time off a manufacturing floor for about an hour to have them listen to a motivational speaker. And you always worry about that because you don't know how much you're going to really get from a motivational speaker. But if I remember right, people were on the edge of their seats—we're talking about older employees, many of them with more than thirty years of experience. They came in with their normal 'I gotta be here; let me sign a list, and I'll get out of here as soon as I can' attitude. What happened was amazing. I watched people that I had difficulty with (and I'm not a bad presenter), and I saw those people getting it. They were laughing; they were entertained, and they seemed to grasp the whole thing."

"It *was* a great two days," I agreed. "I couldn't believe the size of the facility; there was even a fitness center on site. The presentation area was huge too; it held over 400 people. I remember walking from there to the cafeteria, and people coming up to me saying, 'Man, that was the most right-on thing I've heard, and I've been here twenty years' or 'I've been here thirty years.' One of them was a fork truck operator who stopped his truck, got off, shook my hand, got back on, and drove away. One of your assistants, I think that his name was Don Otto, told me, 'That man has not spoken to me in ten years.' There was a huge...I don't know what you'd call it, I mean, there was a buzz...there was this huge buzz happening. People were coming in on day two with a different attitude. They probably heard it was entertaining and all, but there was more happening than just that. I think it's fair to say that we could all feel it. Management people could feel it; you and your staff were hearing lots of good feedback. At each successive session you could see people coming in with a curiosity. They'd heard this was new and innovative; it was something different. But I suppose the most important thing for me wasn't just the buzz and the positive comments, but the injury reductions that the company got, especially in the short term."

"It was amazing," Gary said. "I think we were all floored. We had a very good-performing facility from a safety stand point— one of the top in the country. We did have some injuries, but normally they were pretty minor. After your sessions, we saw— almost overnight—better than a 40 percent reduction in injuries. We didn't maintain that long term, but for three or four months it was amazing. Now, like most motivational presentations, after about three, four, five months, we saw that injury reduction erode a little bit. But it stabilized at 33 percent. It had quite an impact on the employees. When I added up the direct and indirect costs of the injuries over a seven-year period of time, and then factored in the number of employees, we were talking about a potential savings of nearly $7 million."

"You know, I went back to that plant years later," I said, "and there was still a SafeStart card pinned to the bulletin board. I mean, that was a long time after you had left the company."

"I can believe it," Gary replied. "Yeah, I think it was early 1998 when I left. I headed off to be another company's corporate safety director, and then later moved from appliance manufacturing into the automotive industry. I eventually ended up starting my own consulting firm."

"I wish you told me about the $7 million savings earlier," I said. "I could've...."

Gary cut me off, "*Potential* savings. Underline potential."

I resumed my thought, "Well *you* knew what I meant. Now, wasn't there even more potential savings when they looked at off-the-job injuries?"

"At that time the company was self-insured, and you could get data that you can't get today," Gary answered. "But the potential off-the-job injury reduction for the workers who attended your presentation was even higher than the on-the-job reduction. I think the estimated savings were a bit higher over that seven-year period of time because you can't control the medical costs of off-the-job injuries. If we averaged it all out, it was well over $8 million. The potential injury reduction cost saving made it obvious to me that you needed to package what you were doing, so it wasn't a flavor-of-the-month or the latest consultant coming in and doing a motivational speech. We needed to have something more like a course so that the positive effects of the concepts could be extended long term. At that time, you only had one critical error reduction technique: self-triggering on the state to prevent a critical error. Even though you've developed the concepts and techniques much more since then, we still had a tremendous amount of improvement. So I was pretty excited about it. I remember taking the flipchart paper you used—I still have it today—because I knew something good was happening. Although it's creased and torn, I opened it up a few weeks ago, and it got me thinking about those two days."

I smiled. "At that stage, it was just a sort of an awareness thing. The only technique of substance, really, like you said, was to self-trigger on the state. I think it's also very interesting that, potentially, the company could have saved over $15 million. But there was a real limitation to just doing a stand-up 55-minute training session. I remember one of the things you said to me a few months after the sessions. You were telling me about the injury reductions you were seeing and how significant they were, but you were also cautioning me that a lot of this was so dependent on whether I was having a good day. As you said, 'You had two really good days in a row here in Iowa. But you and I both know it doesn't matter who you are. You always have some days that aren't good.'"

I knew Gary was right. I mean, they were two *really* good days; the audience kind of warmed up to me. But, playing off of the audience was something you couldn't always count on, that's for sure.

I continued, "I remember you also told me, 'Don't make the training a one-shot deal, even if it's a full day. Anything that's a one-shot deal will be in one day, and then gone the next. In other words, these are not difficult concepts, but you need to have the people coming back to revisit them. I don't know exactly how many times, but more than once for sure.' So by the time I left your place I decided to start working on developing these concepts into a course in conjunction with the training company who sponsored the safety advisory board meeting down in Dallas."

"There were a lot of decisions to be made: the structure of the course, the media that would be involved, what the training company was going to do, what I was going to do, etc. Eventually we decided they would shoot the videotapes we needed and help market the course, and I would write the workbooks. I remember one phone conversation with a gentleman named Bill Joiner who was in charge of the safety and industrial skills training part of the business. 'Okay, you've got these states leading to these four critical errors,' Bill said. 'And you're encouraging people to be mindful of the states because once you make that error it's too late. You've also been able to show that a presentation of your concepts in front of 2,800 workers led to significant injury reductions.' I told him yes, but I also told him there were a lot of companies where I was implementing these concepts in addition to the traditional, behavior-based safety concepts. These companies got injury reductions much quicker than you would normally get with just the observation-feedback process."

"As we continued talking on the phone, Bill got me to realize that maybe I had stopped a little short of the finish line, and that perhaps there was more involved in the state-to-error risk pattern than I'd thought—something more beneficial for people than just self-triggering on the state. After we talked, I started looking at the state-to-error risk pattern again."

CHAPTER 9—*The Second Critical Error Reduction Technique*

"I know that you took Bill's advice. So where did you go from there? I know SafeStart spread very, very rapidly with lots of success. In fact I even used you a few times to come in and talk in some of those other companies I worked for. But tell me about your development from the point when Bill talked to you. How did you come up with the other critical error reduction techniques?"

"Well, I started working on the course about June of 1997, a few weeks after I got back from Iowa. And by late 1997, I had figured out a few more things that would eventually become the four critical error reduction techniques (CERTs)."

Before going on to the next technique, let me explain something about the first one: self-triggering on the state. As you look at the state-to-error risk pattern (see Figure 12), you see the four states in an arrow that points to the four critical errors also in an arrow. The arrows are pointing to two triangles, a small one and a large one. The triangles indicate that you're going from low risk to a higher risk because once you add the states to the errors,

Figure 12
State-to-Error Risk Pattern

you increase the risk for any activity, whatever it is.

If you look at the first three states—rushing, frustration and fatigue—those are the states you can self-trigger on because they are different states than normal. In other words, you can tell when you're in a rush; you can tell when you're frustrated; you can tell when you're tired. I mean, it takes conscious effort to move or to think or to talk faster than you normally do. You can certainly tell when you're frustrated—hopefully, it's not your normal state. The same thing is true for fatigue. In other words, these states are active ones. You can tell that you are in a rush, frustrated or tired.

And the idea was that instead of thinking about why you were rushing or what was going to happen to you if you were late, get people sort of programmed—rewire their brains so to speak through enough repetition—so that their first thought would be eyes, mind, line-of-fire and balance, traction, grip. In other words, as soon as you realize you were rushing, it would trigger a thought like, "This is a high-risk state. I've got to keep my eyes on task; I've got to keep my mind on task; I've got to be thinking about line-of-fire, and what could cause me to lose my balance, traction, or grip."

I continue on with Gary. "If you were in the fourth state, complacency, the self-triggering technique would be very difficult. Remember, you can tell you're in a rush; you can tell you're tired; you can tell when you're frustrated. But it's not that easy to tell that maybe you've become a little too comfortable with the hazards in your work area or your speed out there on the highway, because it just seems like everything's normal. So there's nothing causing an alarm, nothing raising a red flag, nothing coming on your radar screen—figuratively speaking— that alerts you that anything is different or that danger is present. Complacency causes many problems, but the main problem that complacency causes is that it leads to mind not on task. Or to put it another way, once the fear is no longer preoccupying, your mind can wander—you can start to think about something else."

"For example, think about driving. When you first started driving it took all of your concentration and attention. Things like not over-correcting or keeping the car in the lane took a large percentage of your concentration. You were almost 100 percent mindful of what you were doing. But after a few years of driving, pretty soon you could drive home without even needing to think about what you were doing. However—and this is the key point—when you aren't thinking about what you're doing, your behavior will be what it normally is. You will do what you do normally or automatically or *habitually*. The only way you could change from what you do normally is if you made a conscious effort to change."

"But as I said, since you're not thinking about what you're doing, your behavior will be what it normally is. I'd learned this much from the behavior-based safety process. If you went out and you made enough observations on people looking for critical behaviors, most likely, the people being observed would be more inclined to perform those critical behaviors safely in front of an observer, and as long as you made enough observations, the theory went, eventually some of the critical behaviors would become more natural done safely than unsafely. In other words, they would eventually bring these critical behaviors up to 'habit strength.' So the whole idea with behavior-based safety was to get you doing these things safely enough times that eventually they would become what you're most comfortable doing. The most frequently cited example was the seat belt. Most of us know someone (maybe even ourselves) who started wearing seat belts only when their use became law and increased enforcement brought with it fines. Now you find a lot of people who, even if they took the seat belt law away, would keep wearing it because they're more comfortable wearing it. The key was building things to habit strength."

"What I needed to do was simply flesh out some of the specific safety-related habits that people would need to improve in terms of eyes on task, body position out of the line-of-fire, and balance, traction, or grip. That was the beginning of the next critical error reduction technique I found: 'Work on habits.'"

Gary was looking at the back of the SafeStart card he held in his hand, and then he looked up at me and said, "What do you mean 'the next'? 'Work on habits' is the fourth CERT." He held up the card and then said, "See" (see Figure 13).

"I know it's the last CERT on the card, but I was talking about the order in which I discovered them, which, like I said before, was mostly luck. Is it okay to go on?"

"Yes, you may proceed," Gary said with a smile.

"The importance of having good safety-related habits really came home for me when I was doing some work in early 1996, at a steel mill north of Winnipeg. It was *so* cold out. I remember they set a record for something like 19 days of minus 30 degrees Celsius (-22°F) or below, if you can imagine.[6] It was just a lovely place. Anyway, they don't call it 'Winterpeg' for nothing. Walt, the melt shop manager, and I were doing observations when we

---

[6]Environment Canada, "La Niña—Past Effects of La Niña," <http://www.ec.gc.ca/adsc-cmda /default.asp?lang=En&n=9A2D0685-1>, accessed April 9, 2012.

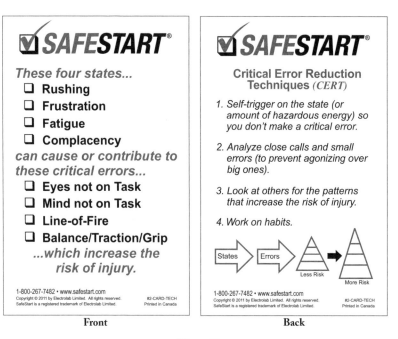

Figure 13
The SafeStart Card

saw a fairly stout, older man pick up a tool off the ground. He didn't, however, have his knees bent; he was lifting with his back. 'How heavy is that thing?' I asked the man. I was just trying to get a handle on how heavy the tool was before talking to him about bending his knees because of the additional strain he was putting on his lower back. The guy was only so thrilled about being observed. He looked at me and said, 'Why don't you pick it up and see for yourself?'"

"I went over. The tool was bigger than a crowbar. When I bent over to pick it up, Walt stopped me right before I grabbed it. 'Larry, just touch it first.' He said, 'Just touch it.' 'Why?' I asked. He said, 'Well, because if you grab it and it's still hot, you'll burn your hand. You've got to learn to get in the habit of just touching things around here because you never know what's hot and what isn't.' I thought, 'That was just brilliant.' You get complacent in a melt shop, and you might not always be thinking that some things are hot and some things aren't hot. Therefore, you need to get in the habit of touching things first and not ever grabbing them."

"So after that, I started adding the specific habits as they related to eyes on task. For instance, we're not just saying, 'Okay, watch what you're doing.' We're also getting people to think about 'Move your eyes before *you* move.' Move your eyes before you move your feet, before you move your hands, before you move your body, or before you move your car. Look before you stick your hand or your foot into something. These would probably be some of the issues you saw a lot of, Gary, with the assembly line or with mechanics, where people would have to stick their hands into places where they couldn't see."

"Yeah," Gary said. "They've done it a thousand times before without getting hurt, so there's a little bit of positive reinforcement of a negative act, in other words, reinforcing a bad habit. They're constantly moving their hands before they're moving their eyes. I've caught myself doing that before, too. So you were working with people to try and get them to develop the habit of moving their eyes first before they moved, whether it was their body or their car?"

"Right. There are some other things that we all know we should do, but not everybody does—at least, not every time. For instance, if you're looking into the sunlight—look twice. The sun *does* play tricks on you. Here's another one: Look around for things that are the same color or texture as the floor because somebody else might not see it with just their peripheral vision, like grey welding rods on a grey shop floor. If you're looking right at them, you see them. But if you're not looking right at them, someone might only realize it when their heel hits the weld rod and they lose their balance. Another habit would be to look at your back foot as you step over things. It's rarely the front foot going over something that makes you trip because you're watching it. Typically, you don't look at your back foot, and you get it caught up on things."

"Here's another habit: Watch where other people are looking. This is especially important if you're riding a motorcycle. Where is the driver looking? Do they see you or are you just assuming that they see you? If you're on a motorcycle and they're in a car, it might be the last assumption that you make. Some other things came to mind too, like if you're in a rush, there's a natural tendency for people to check their watch every 30 seconds. Instead they should be thinking, 'Hey, I'm in a rush; I've got to get to the airport, but checking my watch every 30 seconds isn't going to get me there any faster.'"

"All of the habits I've mentioned so far have been focused on eyes on task. But you've also got habits focused on body position out of the line-of-fire—like keeping your hands out of pinch points—and habits focused on balance, traction, or grip—like not committing your weight to something before you've tested it to make sure it will hold you or that you won't slip. These are all things that people know about because they have learned them through experience. But few people ever try to improve them. It just doesn't happen."

"There are perhaps a number of different ways to explain why people don't try to improve their safety-related habits. But the thing that occurred to me was that adults think that they're safe enough already. This idea was underscored when I was getting people to build their own personal risk pyramid. It was obvious to me from having children that little kids get hurt way more than the adults do. The implication from this fact, however, wasn't so obvious: skills with eyes on task and mind on task were actually learned skills. Each of us learned these skills painfully every day as we banged into things, cut ourselves or fell down. When my wife and I started having children in 1996, we would typically see 15 to 25 visible cuts, bruises, bumps and scrapes on our kids when they were under ten. But if you looked for cuts, bruises and scrapes on my wife and me, you would typically find only one, maybe two."

"So what I realized is that we go from getting hurt about 20 times a week as little kids to about 20 times a year as adults. From 20 a week to 20 a year is a 5,000 percent improvement. So there's a natural tendency for the adults to all think they are safe enough already. Getting people who already think they're safe enough already to put some effort into improving their habits isn't easy. Let's face it, if you thought you were safe enough already because you've improved 5,000 percent from when you were a little kid, the likelihood that you would still be thinking that you've got a problem in terms of personal safety skills isn't very high. In other words, I've never met anybody, Gary, and I doubt that you've ever met anybody who came to one of your classes and said, 'Thank God you're here, Mr. Higbee! I'm just an accident waiting to happen; hopefully, you can help me.'"

"You're right, I haven't," Gary said. "If I had, I'd be a little bit worried about that individual. Safety classes aren't the most exciting things....Okay, where were we? Oh, yes, so you've got two CERTs: self-triggering on the state and developing good safety habits so that when your mind isn't thinking about what you're doing, the habit carries you through. That's two, but what about the other two? When did you discover those?"

"The other two came in 1998, but, like I was saying, there's also this other problem that I had to come to grips with: Nobody thinks *they* have a problem. So as you're moving towards trying to make the concepts efficient for people to use, you also realize that there's another dynamic that's coming into play. Unfortunately, I hadn't quite envisioned that one before I started working on the course. For instance, you could have the best diet in the world, but if somebody thinks they don't need to lose any weight since they're thin enough already, it really wouldn't matter what your diet was or how little it cost because some people might say, 'Hey, I'm trying to gain weight; I'm not trying to lose weight.'"

"So, although nobody was ever trying to get hurt, that doesn't mean the average person is all that motivated to put any further effort—over and above what they'd already decided was enough—into getting hurt less. But I was only beginning to recognize that perspective as our children started getting older, and they started bumping into things. I looked at their shins and thought, 'Wow, that's a lot of bruises,' compared to me or my wife. So now I understood that people and their skills change and get better over time because the pain is really motivating. If you asked, you'll find that 60 to 80 percent of the people in any given room have slammed a finger in a car door once. But that percentage goes way, way down when you start asking how many had slammed a finger in a car door twice. And hardly anybody has slammed a finger in a car door three times. Now, we didn't all quit using automobiles or getting in and out of cars just because we slammed our finger in a car door the first time. Instead, we became a lot more aware of finger/body position out of the line-of-fire because the alternative really hurts. So we have a dynamic that's counter-productive here: If everybody thinks they're safe enough already, how much effort will they put into improving things?"

"That is interesting...and...important," Gary answered, "and it sort of explains why nobody likes safety training. But it doesn't explain how you got the other two techniques."

CHAPTER 10—*The Third and Fourth Critical Error Reduction Techniques*

"So I have the two techniques," I said, "but..."

"You mean you discovered the first and the fourth CERT?" Gary interrupted.

"You're getting a kick out of this number thing, aren't you?" I replied.

"What can I say?" Gary smiled, "I'm a technical guy."

"Okay," I said trying to recover my train of thought. "I also knew that habits and reflexes—alone—don't give you the ability to anticipate a dangerous situation. That's the kind of situation you need your mind for. So I needed to find a way to help people pull their mind back into the 'ball game' from wherever it may have drifted off to. Another one of the things I had learned from the behavior-based safety process was that making observations tended to improve your awareness or mindfulness about safety. The observation engaged you: You had to think about who you were going to observe, which area they were in, the hazards of the area, the personal protective equipment required for the job, the procedures they should be following, etc. By the time you'd done all of that and discussed it with the person you were observing, you were a lot more mindful of safety and what you were doing."

"That's the third CERT you discovered, and it's number three on the card too." Gary replied.

"You aren't going to do this with each CERT, are you?" I asked.

Gary just smiled. "So...observing other people seemed to make the observer more aware or less likely to make an error—just watching other people?"

"Yeah. You probably remember that some of the behavior-based safety firms in the early '90s were actually talking about how the observers were five..."

"Five times safer," Gary broke in. "Sure, I remember that."

"...five times safer. But they didn't make a big deal out of it because they were only concerned with the effect on the person being observed, rather than any effect on the observer. But I never forgot it. So I thought one easy way to keep your mind on task is if you just started looking at other people for these state-to-error patterns. Once you start to see them, you do see them everywhere. In other words, looking for state-to-error patterns tends to pull your mind back into the moment and makes you more aware of what you were doing at the moment."

"So observing other people for that state-to-error pattern will increase my awareness," Gary said. "If I do that enough, then it should make me less likely to repeat the same state-to-error pattern, right?"

"It will do more than that. It will not only make you less likely to make the same error, but it will also make you think more about what you are doing at the moment from a safety or risk perspective. As I said, these patterns are everywhere. I remember years back, one of our consultants asked me, 'Do you ever stop thinking about this stuff?' I told him, 'Shayne, I would—but it's everywhere. I mean, just when you want to stop thinking about it, some waiter spills a tray of drinks and things come crashing down, and it brings you back.'"

"That reminds me, remember Puerto Rico?" Gary just nodded and smiled.

"I think that's one of the funniest and most ironic examples of how these state-to-error risk patterns are everywhere. It was shortly after you came to work for us part-time in 2002. We were both speaking at an OSHA conference in Puerto Rico. Although it's kind of odd, Puerto Rico and the Virgin Islands are in OSHA Region 2 along with New York and New Jersey. I think this had to do a lot with immigration in the old days, people coming from Puerto Rico to New York. So, the OSHA regional administrator from New York also attended. One evening she was having dinner with us, and I recall her saying something about (I'd heard this before and was sure you had, too) 'I don't believe in behavior-based safety. I believe in hazard reduction and elimination.' And we're sitting there..."

"Oh, yeah," Gary interrupted, "you're right. It was New York, and she was the head of that region. We were sitting at a round table, and our waiter came in with a whole tray full of drinks, bumped into a waitress and almost dumped it right on top of her."

"The whole tray of drinks came crashing down beside her," I continued. "I mean, it only missed her by inches. I remember looking at her, then looking at you and trying to keep from laughing. Then I think I said something like, 'If she was wearing that whole tray of drinks right now, I wonder what she would be saying about behavior versus hazard elimination?'"

"Yeah, and it was interesting because even when that happened, I don't think she ever got it. All she was thinking about was getting rid of the hazards. But a tray of drinks is a hazard that's really hard to engineer out," Gary said. "You have to rely on the wait staff not running into each other. But in this case neither one of them was watching where they were going, and they crashed. Frankly, I've seen that happen in quite a few restaurants. In fact, at some restaurants, it gets to be a joke because so many plates and meals get dropped in a night that eventually everybody starts clapping when it happens."

"So what you find," I continued, "is that while you might not see these state-to-error risk patterns every second or every minute, you'll certainly see them every hour or every day, which means you can rely pretty heavily on being able to see other people in these state-to-error risk patterns to help you fight complacency. Obviously, if you're out by yourself in the middle of nowhere, you'll have to recognize these patterns within yourself. However, most of the time we are in a setting with lots of other people around, like in rush-hour traffic. In those kinds of situations—crowded airports, shopping malls around Christmas, etc.—you will see state-to-error patterns literally every minute."

"Gary, weren't you the one who told me a story about a guy with his Christmas list in a shopping mall?"

"Oh, that was a great one!" Gary said. "This guy was a typical male shopper, waiting till the last minute. He wasn't really shopping, he was *buying*. He was going from store to store; he had bags under each arm; he had his list in his hand. I watched this guy as he walked down the aisle in the shopping mall with his eyes glued to the list. A few feet in front of him was a huge support column; it had to be about four feet (one meter) in diameter. He was walking fast as he could, and he walked right into it, stumbled back and said, 'Excuse me.'"

"I couldn't believe it. He didn't even realize it was a pillar that he hit; he ran right into it. The guy could have really used SafeStart. He was definitely rushing around trying to multi-task. I didn't see him long enough to determine whether he was tired or frustrated before the collision, but with all crowds and hassle, it wouldn't have surprised me if he was. He was probably also a little complacent. And look what happened: He wasn't thinking about what he was doing; he didn't have his eyes on where he was going, and he walked right into a pillar. Now, he could have made those same errors and if the pillar hadn't been there, nothing would have happened. But this time a pillar was there, and he didn't even see it. It probably wasn't the most professional thing to do, but I couldn't help laughing. He did look a little frustrated when he looked at me and saw me laughing. But it was amazing to me that somebody would bang into a support column and then would say, 'Excuse me.'"

"It's a pretty funny story," I said. "I've also seen a YouTube clip where a lady was texting and walked right into a fountain."

"Okay, moving on...that's three CERTs," Gary said. "What about the last one?"

"Aren't you going to say it?" I wondered.

"What do you mean?...Oh, the number thing. No, I'm over that," Gary smiled, but his tone was back to business.

"The last CERT is sort of a combination of the other three. It's the 'how do you get continually better at this?' technique. You see, we get a lot of free learning experiences, and maybe because they're free that's part of the problem: Nothing happened, so it's no big deal. But we have an awful lot of close calls and make a lot of very minor errors. Take, for example, a momentary loss of balance. It happens about once a day, or at least once a week, but rarely do adults stop and think about it. It's like that quote by Winston Churchill, 'Men occasionally stumble over the truth, but most of them pick themselves up and hurry off as if nothing had happened.'"[7]

"The same thing is kind of true in a literal sense when you just momentarily lose your balance. Instead of asking yourself, 'Hey, what was going on there?' the normal response is more like, 'Oh, that's weird; I don't normally do that,' and then they carry on. Or consider somebody who turns without looking and bumps into something, and they say, 'Hey, I don't normally do that.' However, the reality is yes, they do normally do that, but normally, there isn't a filing cabinet in the way. That's what was different."

"That's what the last CERT is all about: getting people to pause and analyze the close call or the small error. The first thing is to just to ask: 'How could it have been worse?' or 'How easily could it have actually been worse?' Asking these questions would help to fight complacency from a motivational point of view. When you start thinking, 'Hey, that could have been it; that could have been game over right there if that truck had hit me,' hopefully, those kinds of thoughts will help fight complacency a bit. But the second part of this technique would be to keep it from happening again. So whenever you receive a minor injury (you bump, bang or scrape into something) or even have a minor close call (you lose your balance or a horn honks at you because you didn't see the car in that lane), we want you to ask: 'Okay, was this a state like rushing, frustration, or fatigue that I didn't self-trigger on quickly enough? Or was it complacency leading to mind not on task?' If it was complacency leading to mind not on task, then it's either a safety-related habit that still needs more work, or you need to put more effort into observing other people for those state-to-error risk patterns. However, most of the time what you'll find is that it was actually one of those safety-related habits that still needs more work. So this technique of analyzing close calls uses a combination of the other three techniques, so that when something did slip through the cracks—hopefully a minor thing, something that was virtually free—you would be able to analyze

[7]Winston Churchill, "Quotes," < http://www.brainyquote.com/quotes/quotes/w/winstonchu135270.html>, accessed April 9, 2012.

it, and learn how to prevent it, or similar things, from happening in the future. In summary: Was it a state I didn't self-trigger on quickly enough? Or if it was complacency leading to mind not on task, then most likely it was one of those safety-related habits that still needs a bit more work."

"If we could get people to put a bit of effort into these four critical error reduction techniques, then we could take a huge bite out of the injuries that occur in the 'self' area, which I already knew were well over 95 percent of the injuries on and off the job. When I presented these concepts in Iowa in 1997, I thought I just needed to make people aware of the state-to-error risk pattern and to become more mindful of the state. But by January 1999, when the course was complete, I realized that the challenge really was motivating people to put some real effort into getting better at these four techniques. As mentioned before, I had already figured out the mindset that was going to be counterproductive to this, because all the adults in the world have concluded that they're safe enough already. So you don't find people running into safety training classes, making sure they had a good seat at the front of the room, notepads and pens ready just in case there was something said that would be critical to their personal safety. You find exactly the opposite. You find that getting people to safety training is a heck of a lot of work. It's like herding cats, and the managers and the supervisors are the worst. They are like professional dodgeball players. It's almost impossible to get them to go to safety training classes, let alone go first."

"So now, after all these years and all this effort asking thousands and thousands of people questions about how they've been hurt, we've finally got four states leading to four errors, and we've got four critical error reduction techniques to deal with that state-to-error risk pattern. We've finally got things into an efficient enough package that is easy enough for anybody to understand. But what we also found out is that nobody really wants it, at least not for themselves, because everybody thinks they're already safe enough already. However, even though people don't think they have a personal safety problem, the companies they worked for were not necessarily of the same mind. They weren't happy with their injury rates staying where they were. Some of them wanted to get better; some of them believed in continually trying to get better. A lot of them actually had a goal like 'target zero,' where they weren't going to rest until all of the injuries had been prevented."

Even though there are some counterproductive dynamics happening with people on a personal level, there's also concern for the employees at the corporate level, which meant that even if the employees didn't want to go to safety training, the companies were going to insist. We had companies buying this training course for their employees but, interestingly enough, sometimes not even safety people were attending the sessions for themselves. I know because I would ask at public workshops, 'Did you come here to evaluate this course, these concepts and techniques, to see if it might help out people you work with or people who report to you?' And everybody in the room would put their hand up. Then I'd ask, 'How many of you came here today because you were personally fearful for your own safety and you were hoping that these concepts and techniques might save your life?' Now nobody put their hand up, and quite a few people even laughed. And then I would look at them and say, 'So, you came here to see if you could find something to help the lesser mortals at your company, didn't you?' They would all kind of laugh sheepishly. And I would look at them and ask, 'Do any of *them* think they're lesser mortals?' They'd shake their heads.

"In spite of the 'I'm safe enough already' attitude among employees, and the 'I'm too busy to attend safety training' attitude among managers and supervisors, the corporate drive to reduce injuries and improve safety performance, reduce costs, improve productivity and efficiency, and reduce downtime and accidental equipment damage really helped the rapid spread of SafeStart across North America. I'm sure you saw some of the same things at the various corporations you've worked at. You even got some of them to implement SafeStart, and saw something like a 50 or 60 percent decrease in injuries."

"I didn't get to do it at all of the sites," Gary said, "because a lot of these corporations were decentralized. Each individual site was autonomous, so they could choose what they were going to do. Of course, I was still doing all the traditional safety compliance plus the traditional behavior-based safety observations, so there were lots of activities going on. But those sites where we implemented SafeStart had some significant injury reductions. I was starting to look at safety a little bit differently. Instead of approaching it solely from a compliance standpoint, I was starting to focus more on achieving injury reductions. You see, unfortunately, there's a point in time where compliance and injury reduction split. I was struggling with that and moving from one plant to another trying to get people to look at things differently. I had a lot of success at some of the individual units with 50, 60, 70 percent injury reductions, but I couldn't always get SafeStart to the rest of the corporation. It was frustrating; it was almost like they really didn't want to reduce their injuries. They just seemed content with compliance."

"I talked with you at one of those companies about some of these frustrations," I said. "I remember asking, 'If you were to add them up, how many injuries realistically do you think you've helped to prevent? I know you didn't prevent them all by yourself, but about how many injuries have you helped to prevent during your career? You know, do the math for all of the companies where you worked, based on the injury reductions that you achieved over the years...'"

I sort of had it in my head—without knowing the exact numbers—that Gary had probably helped to prevent somewhere in the neighborhood of 6,000 to 8,000 injuries, or maybe as high as 10,000, in the 30-plus years that he had worked as a safety professional.

"I did the number crunching," Gary said, "and I went through the process. I think you're probably giving me credit for more than I deserve. I was only one of the players that had an impact. There were lots of other people that were involved. I remember you thought it might be as high as 10,000, but that would have been a real stretch."

"By 2002," I continued, "we had already implemented the SafeStart concepts and techniques with around 300,000 people. The numbers were growing rapidly. We were achieving significant injury reductions of about 50 percent even at some of the places that did a half-hearted job of implementing it. I had heard you talk about the friend who was killed in one of your facilities, so I knew that safety was more than just a job for you. I can't remember exactly how I said it, but I think it was something along the lines of, 'You can keep working at those companies, and you can probably prevent another 5,000, 6,000, maybe even 10,000 injuries, tops, or you can come and work with us, and maybe help to prevent millions of injuries, because SafeStart isn't going to just stay in Canada and the United States.'"

Gary looked out the window and paused—a thought about his deceased friend perhaps flashing through his mind. Then he said, "Yeah, I thought about your offer for a while. It kind of coincided with my thinking at the time that I didn't want to focus on compliance anymore because you can be totally compliant and still have serious injuries. Instead, I wanted to work on injury prevention. I was also looking beyond the workplace and thinking about injury reductions at home, in the community and on the road. So it was a no-brainer when you offered me a position. Of course, I would have rather you offered me royalties, given that it was all *my* idea to produce a training course. But it's been a great run. And so, yeah, in 2002, I think that's when I made the switch and got out of corporate safety jobs. And the fun hasn't stopped since."

CHAPTER 11—*Motivation: The Need to Get the Family Involved*

So Gary formally joined the team in 2002, right around the time SafeStart went international. That happened when Coastal Training Technologies (now known as DuPont Sustainable Solutions) picked up SafeStart for Latin America. At that time they only wanted Mexico and Latin America, and we started working with them to sell SafeStart internationally. We translated SafeStart into Spanish, and then into Portuguese. We had a little bit happening in Europe because Fiat had translated the program into Italian. We also had the program available in French because of the Quebec market in Canada. So we had English, French, Spanish, Portuguese, and Italian—five languages—and the program had been on the market for only three or four years; things were certainly moving along.

We also had been so successful with companies reducing their injuries that we were now offering them a guarantee—a guaranteed amount of injury reductions. It was the fact that we were willing to put our money where our mouth was that got us the job at the U.S. Postal

Service's General Mail Facility in Denver where about 2,800 people worked. We had guaranteed them a 35 percent decrease in recordable injuries. The only caveat was the people had to go to the Employee Overview sessions Gary and I were conducting, and then complete the rest of the course. The overview sessions would take them through to the end of Unit 2. Then they would have to attend Units 3, 4 and 5 conducted by USPS personnel to be considered an eligible person from the standpoint of the guarantee. In other words, we weren't guaranteeing injury reductions for people who hadn't taken the course, but that was the only caveat. We had one year to show a 35 percent reduction in recordable injuries. This shouldn't have been a problem since hardly any company we'd worked with had achieved less than a 35 percent improvement. But I do recall Tony Ploughe the Safety Manager saying to me, "Are you sure you're gonna be all right with this? Our folks are pretty tough."

I didn't know what he meant by this. I mean, I knew about the disgruntled post office employee who had shot 14 of his co-workers and had heard the jokes about "going postal," but I hadn't really considered the possibility that the culture there might be a lot different than anything I had ever run into. When Tony said, "Our folks are pretty tough," I was thinking, "Tough? Tough compared to pipe lining north of the Arctic Circle when it's 40 below or tough compared to hand logging on the Pacific Northwest where it's rainy, it's cold and it's miserable; you're on a 45-degree hill, and you're using a 3-foot chain saw to cut down 200-foot-tall Douglas fir?" So when he said, "Our folks are tough," I'm thinking, "Bring it on."

Gary also had lots of experience with tough union environments, including the automotive world, so I thought he wouldn't have a problem with this, either. I'm handling four days; he's going to handle the other three days—the weekends. We've got a month to deliver the Employee Overview sessions, plus a Trainer Certification and a Leadership Support session. But the majority of the work was going to be presenting these half-day overview sessions to the employees.

I found out what Tony meant when I got there for the first day's session. When he said, "Our folks are tough," he didn't mean they're tough like a bunch of Hell's Angels. What he was trying to say was, "Our folks are tough" from an apathy, couldn't-care-less point of view. If their body language could have talked, here's what it would say, "We're not going to pay any attention to you. You can stand up there and talk all you want, but we're going to sit here; we're going to read magazines; we're going to talk to each other; we're not going to read your workbooks; we're not going to do anything you tell us to; we're not going to really pay attention to you if we don't want to. Sure, we have to be here, but that's all they can make us do. And we're going to take the pencils at the end of the class because we deserve at least that much for our time...." It was much different, much worse than any class or series of classes at any company I had ever done in any industry—it was brutal!

Tony did not introduce me as somebody who had come up with innovative concepts and techniques. He introduced me as somebody who used to be on the soap opera "The Young and Restless." He said, "They could care less for the safety stuff, but this might help, Larry, 'cause they all watch soap operas." It was so bad that I remember phoning Gary two or three times a day, sometimes four times a day, saying, "Gary, you've never seen anything like this! I've never seen anything like it! It's apathy gone wild." I told Gary about the people reading the magazines, about the people sleeping, about the people who wouldn't read the workbooks, and about the people who would talk to their co-workers to keep them from reading the workbooks even if they wanted to. I told him how difficult it was to keep the class together. I mean, it took everything I had. I would have walked away from the post office deal if I could have, but we'd guaranteed them a 35 percent decrease in injuries or they got their money back. (At the time this was a significant amount of money, so we couldn't just walk away.)

"Gary, do you remember me phoning you three or four times a day from Denver and telling you how bad it was? When I told you not to talk about your safety credentials, but to talk about how you got your purple heart, you probably thought I'd lost it."

"I thought you were over the top," Gary said. "I thought to myself, 'How many times is he going to call me? It can't be that bad.' I had been in tough spots many, many times before, but when I finally got there—well, frankly I thought, 'You low-balled me.' It was worse than I ever expected. I had one woman come into the classroom, sign the attendance sheet, go to the back of the classroom, lie down, pull her parka over her head and go to sleep. I asked Ernie, the safety guy there, 'Is there anything we can do? I mean, how am I going to train her? She's asleep.' And he said, 'All we require is their attendance. We can't really require them to be listening.' I thought, 'How is this ever going to work?'"

"I think it was the very next week that I was there, right after your first weekend, that I started getting desperate," I chimed in. "Like a drowning man grasping at straws, I started trying anything that I thought might work. On the second day that week, I began telling people the course was available on the Internet free of charge to any of their immediate family members because the post office had bought SafeStart, 'So if you want, you can take SafeStart home to your families.' It was just something I threw out. I was using everything I had at this stage of the game, and it didn't look like anything was going to work, so I just threw that out there."

"I had barely said the words when there was a pause in the class. The people, who I didn't think heard a word I'd said, finally started looking at me and paying attention. The more I talked about the online course, the more they listened. People started to wake other people up. I remember watching this one man wake a woman up. She looked at him first, and I couldn't see what she said, but it looked like she was telling him something like, 'Bug off. Leave me alone.' He looked right back at her and said, 'No, you need to hear this. This is something for your children.' She sat up and started paying attention. I don't know why it took me so long to figure this out, but from then on I started asking people, 'How many of you care more about your family's safety than your own?' Everyone put their hand up."

"I knew enough about adult learning theory to know that teaching has the highest retention. In other words, if you get somebody to turn around and teach what you've just taught them, they're going to retain what they learned much longer. This holds true even if the person has little interest in learning the material. Unfortunately, when it comes to personal safety, nobody thinks they need to learn anything at all because they all think they're safe enough already. Talking about the online course turned the situation around completely. It also gave us invaluable feedback. What happened is that people would come back and say, 'Hey, I tried to get my kid to look at your online course, but they didn't want to do it. So, I was wondering if you could make a video to help motivate them to *want* to take the course online.'"

"As it turned out, this was the beginning of *SafeStart Home*. We started off just trying to make a short video that would motivate people to take the Internet-based online course. But working with our Latin America distributor changed our initial direction. At this time, even a low-speed Internet (forget about a high-speed one) wasn't available in a lot of these places, and there was no way of knowing how long it would take to get it there. So instead of just making a short video that would motivate kids to go to the online course, we decided to make a mini version of the course, but we made it more video-based than workbook-based. That was our first DVD, *Taking SafeStart Home*. It had three parts, about 57 minutes in total, with approximately 19 minutes for each of the parts. The DVD would take kids through a shorter version of the basic, five-unit SafeStart course. But it also included telling their own personal SafeStart stories, as well as analyzing the stories for states and errors, and figuring out which critical error reduction technique might have helped to prevent the injury."

"What we started to see—almost immediately—was that with these DVDs for the home, we had moved into an area of safety that employees truly cared about. While it was true that the adults all thought they were safe enough already and were only so interested in improving their own personal safety, they were *very* interested in the safety of the people that they cared about and were responsible for—their children. Just like companies were concerned enough for their employees to buy SafeStart and train them, you could also find the same dynamic occurring with parents and their children. They didn't think *they* had a problem, but they thought their *kids* could benefit from the concepts and the four critical error reduction techniques. Now when we presented the course, we started to see a much, much more interested group of adults."

"After the post office experience, we changed the way we presented our materials in class. Instead of just having the SafeStart binder with the five question-and-answer workbooks in front of them, we put the *Taking SafeStart Home* DVD on top of the binder. So the first thing people saw was this DVD for the home. We wanted them to realize right up front that this course was much different than a workplace safety compliance course about confined space entry, hazard communication or bloodborne pathogens. Employees started seeing value—not merely for themselves—but they started seeing the value for the people that they cared about. Just like companies were worried and concerned enough about their employees getting hurt that they were willing to make some effort, and spend some time and money improving safety for their employees, parents had the same type of concern and motivation for their children."

"*Taking SafeStart Home* was the first program." The next one we produced was one for parents. This was called *Hurt At Home.* Then we made one for teenage drivers, and we upgraded the online course. In the process of making all of these programs for the family, I became aware of an incredibly shocking statistic: Just in Canada and the United States, there are over 7 million unintentional injuries to children requiring a visit to the emergency department, and somewhere between 7,000 and 8,000 children die accidentally every year. That represents over 180 percent of all the workplace fatalities put together.[8] You know, it's sad when a 48-year old millwright dies; nobody's saying it isn't. But that person had most of their life behind them. If an eight-year-old dies, they never really had a chance. I'd heard lots of people say, 'Kids have to get hurt, that's how they learn.' But if you think about those 7,000 to 8,000 kids a year, how much are those kids learning? So what started out as something simply to help improve the training process evolved into, I don't want to say a campaign, but it became much, much bigger and much more important than just workplace safety."

When I finished, Gary said, "While the basic concepts and techniques of SafeStart haven't changed all that much, adding the *SafeStart Home* component so employees can take care of their families has really changed their perspective. The same thing happened to us when we took SafeStart overseas. It wasn't that our technology was wrong or the concepts wouldn't work. Actually, the technology didn't change whether you worked in China, or South America or wherever it was, but their culture made them see things differently. It was really hard to get the resources—time and money—because they weren't thinking

---

[8] These statistics represent our analysis of the situation of children from birth to age 15. They are based on an extrapolation of the data found in National Safety Council, *Injury Facts, 2011 Edition*, Itasca, IL, pp. 11, 12 and 29. The data is from 2008.

CHAPTER 12—*Going Global: The Challenges in Other Cultures*

"I think it would be fair to say that in some of these countries where we've introduced SafeStart, the level of concern that companies had for their employees wasn't anywhere near what it was in North America," I said. "Now you might argue that North American companies really weren't motivated by care and concern; it was just litigation and liability, pure dollars and cents. I'm not saying that wasn't part of it because, obviously, as you know, the lawyers have a field day with gross negligence cases; it's very expensive to be found guilty in the United States. So I'm not suggesting that there is no monetary or financial motivation. But there is also a real interest in not harming employees and not harming the environment that is sincerely felt by a lot of companies in Canada and the United States. But this motivation was not nearly as strong in some other places where labor is overly abundant. Deaths in some of these places only cost a company $1,000. So I agree with you, Gary, it was the time and resources for safety that culturally hadn't been something a lot of these companies had ever come to grips with or gotten their hands around, because they didn't have to."

"I recall the first time I was flying over the Pacific Ocean to Asia to do a Trainer Certification session for a French multinational mining and ceramics company. I remember thinking, 'What if it isn't the same four states or the same four errors in China, Thailand, Korea, Japan or Malaysia? I mean, what if it's a completely different dynamic that gets people hurt in these cultures?' I remember lifting up the window shade of the plane, looking down at the Pacific and saying to myself, 'You know, this isn't the best time to be thinking about this. You should have thought about this *before* you got on the plane. If it isn't the same four states and same four errors in Asia, this is going to be the longest three-day Trainer Certification session you have ever done.'"

"I'm just wondering," I asked Gary, "did you have some of the same trepidation as you were heading over to Asia for the first time?"

"You'd better believe it," he said. "You never know how another culture is going to perceive SafeStart. But one of the things that encouraged me was the fact that all of these cultures know about driving, and I was pretty confident that SafeStart would work anywhere they drive cars or other vehicles. So when I went to China, I found that the Chinese have no problem with understanding the states and the errors. The idioms might be a little tricky, and you might have to explain a few things. For instance, line-of-fire might have to be explained in a little bit different way, but they still get the states and they still get the errors. Now that I've had three trips to Asia with a fourth one coming up to China, I don't expect to have any trouble this time. You've got one coming up to China pretty soon, too, don't you?"

"Yeah. I'll also be going to Malaysia and Singapore. It is interesting, though, that it's the same four states and the same four errors everywhere. It just makes sense because either hazardous energy hits you—line-of-fire—or you move into it. Of course, you wouldn't move into it if you could see it or you were thinking about it, unless you couldn't stop yourself from moving into it in the first place, in other words, you lost your balance, traction or grip. So it's not surprising that it's the same four errors because nobody, anywhere, ever wants to get hurt."

"What I have noticed travelling around the world, though, is that the relative importance of the SafeStart state varies with the culture. For example, when you ask a North American audience, 'Which is the worse state for you, or which state has caused you the most grief, in terms of injuries?' usually about 40 to 50 percent of the people will put their hands up for rushing. The other three states are usually split equally, sometimes with complacency favored a little higher. But when you go to other places, like Latin America, usually 40 to 50 percent pick complacency as being the worst. There you'll find that rushing isn't as big a problem. If you've travelled around Latin America, you may have noticed just how difficult it is to get a club sandwich in a hurry or how difficult it is to get the people at the restaurant to rush. But yet, at the same time, you'll see rampant rushing on highways and in cities. Driving is the last place where you want everybody to be rushing. In Sao Paulo, you'll see people on motorcycles riding the dashed lane lines between slower-moving cars and trucks. The people on the motorcycles in Sao Paulo or some of these cities are going 50 or 60 miles an hour (80 to 97 km/h). The riders know that not everybody signals when they move over or change lanes. So you ask, 'How many motorcyclists die per day?' And they look at you and say, 'Three to five per day.' So it is the same four states, the same four errors and the same four critical error reduction techniques all over the world. The ratio of rushing compared to the other states may vary, but basically it was the same state-to-error risk pattern everywhere."

CHAPTER 13—*Extending the Concepts to Decision-Making*

"Okay," Gary said, "so now you've taken the basic concepts and techniques to most of the world, or at least to the G20, but there also have been some further developments with SafeStart beyond the creation of *SafeStart Home*. You've extended the application of the basic concepts beyond the four critical error reduction techniques, haven't you? Perhaps you should talk about that."

"Well, I said, "we initially wanted to position SafeStart—and this is more from a business or marketing point of view rather than a methodology point of view—as being different from the other behavior-based safety offerings out there that focused on deliberate risk. We wanted SafeStart to be primarily about unintentional risk, focusing on the errors and mistakes people never wanted to make in the first place. Since it wasn't about deliberate risk, SafeStart didn't have a discipline issue, which was a pesky problem with behavioral-based safety programs, especially the supervisor to employee observations we talked about earlier. So we positioned SafeStart as being about unintentional risk, about error. That made it different from other behavior-based safety programs."

"However, I don't think you, me or any of our SafeStart consultants, or for that matter any safety professional worth his or her salt, would deny that those four states did not also influence decision-making or the deliberate risk component that is quite often present in the injury/incident equation."

Gary looked at me as if to say, "Where did that come from?"

"Okay," I said, "let me try again. Because every injury is unintentional, there is always an unexpected event that initiated the chain reaction that led to the injury. Once these contributing factors are lined up, the theory was that as the last domino fell, the consequence was just a matter of luck combined with the overall amount of hazardous energy. So, the methodology in safety was to stop that chain reaction."

"Normally, any one of those factors—like you said about the incident that involved your friend—if any one of those seven or eight contributing factors had not occurred, that chain reaction would have stopped. So that was the methodology and the thinking. But many times there was also a deliberate risk factor involved. In other words, one of those dominoes in the equation was actually something that somebody did on purpose."

"For instance, let's look at something simple, like jaywalking. Everybody knows that when you look and there are no cars coming in either direction, you can walk across the street without getting hit. So, yes, you can wait for the light, or you can walk across the street. Now, somebody could say, 'Well, you're deliberating putting yourself in the line-of-fire,' and you're thinking, 'Yes, I know, but there aren't any cars coming.' And they said, 'Well, what if you slipped and fell, hit your head, and you were lying unconscious on the ground in the middle of the road when a car came along?' And you could look at them and say, 'Yeah, well, what if a meteorite hit me on the head and knocked me unconscious too?' There are a lot of *what-ifs* in this world, and only so many are likely. So if you don't see a car coming in either direction, then a lot of people are going to cross the road. I mean, it's not like we've never done it before."

"All I'm getting at here is that quite often (I knew this from my own personal experience) there would be an additional risk factor involved—a piece of personal protective equipment not employed or a procedure not used—but it would be intentional. The reason the guy didn't lock and tag it out wasn't because he didn't know how to, it was because he did not believe somebody else would turn the power on when he was working on the machine. Because decisions appear so easy to reverse, there was a huge focus, almost a singular focus, in safety on all of this deliberate risk. As a result, most people believed that it was deliberate risk that caused all of the injuries. That was the way everybody, including the behavior-based safety people, talked about injury causation. But I looked at the situation and thought, 'Yeah, but nobody is ever trying to get hurt. There still has to be a triggering event, and more than 95 percent of the time, the triggering event is a critical error or combination of critical errors.'"

"However, that didn't mean that I was totally oblivious to deliberate risk factors. Nobody was. To suggest that rushing, frustration, fatigue and complacency do not influence decision-making would be ludicrous. I mean, I don't think anybody who has any experience in safety, especially the industrial safety field, would ever say that rushing, frustration or fatigue are not contributing factors to errors and injuries. We all know, personally, that they contribute to our decision-making or poor decision-making."

I knew Gary had a great story about how the four states could influence your decision-making, even if all you were doing was taking down Christmas decorations. So, I asked him if he'd mind telling it.

Gary looked at me and said, "Well...all right. But this wasn't what I had in mind. However, it's a good story with a good message, so I guess I might as well. And you're right, it is interesting that even when we know how to do things, we don't *always* do them correctly."

"So, it was a January day, about the middle of the month. My wife, Jan, loves to put up Christmas decorations. At that time we had a two-story home and had a lot of Christmas decorations up. It wasn't quite like the movie 'Christmas Vacation,' but it was getting pretty close. She also wanted the Christmas decorations taken down right after Christmas. You're not allowed to turn on the lights until after Thanksgiving, and you can't turn on the lights after New Year's. If she had her way, she'd have me take them down on the second of January. Normally, living in the Midwest means we don't have much of an opportunity to do that because it's cold and there's snow on the roof. That's what happens a lot. But this particular year, we had a thaw about the middle of January. It was a really, really nice warm day, so I decided that I'd ask my boss if I could take half a day off, run home in the middle of afternoon and start taking down the Christmas decorations. Since the garage was attached to the house, I would put a ladder up on the garage, climb up onto the garage roof, and then because I didn't have another ladder, I would pull that extension ladder up, straddle the peak of the garage, extend the ladder up to the second story roof and just climb up there. Then when I got up there, I'd tie it off—I had a hook there to tie off to—get up on the roof, and start removing the strings of lights, the reindeer and all the other assorted things that we had up there. I would come over to the ladder, climb down the ladder and put down the decorations I had taken off the roof. Because it was quite a load, what I'd do is go back up, untie the ladder, take the ladder down, extend it from the garage down to the cement driveway, take the lights down, carry them into the garage, and hang them up or put them away in storage. Then I'd repeat the whole process. Back up to the garage, back up again to the top of the house, and then I'd take more lights down. Finally, for some reason, I'm guessing I was a bit complacent, and I was in a bit of a hurry, all this going back up to untie the ladder, then down, then up—did I mention I was tired and in a bit of a rush...? Anyhow, I wasn't tying the ladder off any more because that way I didn't have to go back up and untie it. You know, I'm a safety professional; I know I ought to; most of the neighbors wouldn't tie it off to begin with, but..."

"A *certified* safety professional," I interjected. "So, you didn't have any issue with tying off 100 percent of the time in the workplace where you were corporate safety director, right?"

"Well, no, that was the rule," Gary admitted.

"You didn't have a policy," I chided, "about tying off the ladder for the first hour, but after that you didn't have to tie the ladder off any more because it's too much of a pain, and it takes too long."

"Right, we didn't allow that exemption at work. We only allowed it, evidently, at my house."

"Well, that's just it though, Gary," I said in a 'matter of fact' tone of voice. "That exemption isn't allowed just at your house. It's allowed at everybody's house once the level of rushing, frustration, fatigue and complacency gets high enough for that person."

"So what you're saying is that I accepted some additional risk because thought I could climb the ladder safely?" Gary asked.

"Right. You weren't really afraid of anything happening with that ladder. It's a personal thing, the level of complacency. Complacency has to hit a certain level before it starts to affect decision-making. What people get confused with, I think, is that they believe the level of complacency is an absolute thing. But it's never absolute; it's like a sliding scale where there is going to be an increasing amount of complacency the more often you do something and nothing bad happens. You've referred to this in many sessions as the positive reinforcement of a negative act.

But what this story does—and I know I'm interrupting—what this story does for me is that it illustrates how, regardless of the knowledge and the experience of the person, there is a level of complacency that everybody can get to that will affect their decision-making and their willingness to take deliberate risks. And if you add some frustration to the mix—or some rushing or fatigue—things only get worse, in terms of the decision-making."

"I told you I didn't want to tell the story, and now I'm getting a lecture," Gary complained. "But, you know what? It's a good story, and it illustrates some things so...where was I? Oh yes...I continued to go up and down without tying off the ladder. All the going up and down made me hot, so I took off my jacket, and eventually I took off my shirt. Now, I only had a heavy t-shirt on. It was near the end of the process of going up and down and up and down, when I was up on the top level of the roof, I heard a noise. The noise was the ladder falling off the garage roof. You see, you don't just tie it off to protect yourself when you're going up and down. It might also be a pretty good idea to tie it off so that the ladder doesn't fall, because then you have no way to get down."

"So now I'm on the roof of the house; it's getting closer to sundown; the sun's starting to disappear behind the trees, and it's getting decidedly colder. My jacket and my shirt are on the garage roof, not on the roof that I'm on, and I'm trying to figure out how the heck I'm going to get down. There was absolutely no way I was going to leap from the house roof to the ground, not in a million years. I'm sitting on the roof, thinking, watching the cars slowly pass. The house was located on a lane, so as people drove by I'd wave at them because that's what I always did. They were nice enough to wave back at me; they probably thought there was nothing wrong."

"So there I am. I can't get down, and it dawns on me that my wife's at work. She'd planned to work late that night and wasn't going to be home until somewhere around 8 o'clock. By this time, I'm getting cold. Actually, I'm getting *very* cold. So I start to think, and an idea begins to take shape. Maybe I could hang down over the peak with one hand on each side of it. I'd be fairly close, and maybe I could drop down on to the garage roof. But looking at that drop and thinking about my age at the time and realizing I wasn't the athlete I used to be, that idea left my mind pretty quickly."

"But as the temperature continued to get colder, and as I continued to be frustrated and, to some extent, angry at myself for what had happened, I decided that maybe hanging from the house roof wasn't such a bad idea after all. So I went to the edge of the roof again and looked down. It was quite a ways from the peaked roof of the second story down to the peaked roof of the garage. The pitch of the garage roof wasn't the same pitch as the house. So it wasn't just one story down, it was a little bit more. But I'm very cold, so I decide to try out my idea. I lean out over the roof, and I end up holding onto the roof with one hand on one side of the peak and the other hand on the other side of the peak. As I looked down, I decided that my idea wasn't going to work. Unfortunately, with the weight I'd gained, there was no way I was going to be able to pull myself back up to the upper roof. So now it wasn't a matter of whether I was going to drop down onto roof of the garage, it was how long I could hang on and when I was going to drop."

"So I decided to do some planning. It was a heck of a time to start planning things, but your decision-making process isn't very good when you're cold and frustrated. So I decided that maybe dropping straight down wasn't a very good idea because quite likely one leg would go east and the other would go west, and that's not going to be a very good deal. To make this work I decided that I'm going to have to swing to one side of the garage or the other, and I'm going to have to push off the wall to have enough space to twist around and land properly. Now to make matters worse, there was a window underneath the peak, so now I'm going to have to swing back and forth a little bit further because I can't push off against the window."

"After deciding on a course of action, I started to swing back and forth. You see, my plan was to hang on with the left hand as I swung to my left, let go of the right hand, use my left foot to push off the wall and then land—like a cat—on the roof. Unfortunately, this cat didn't land on his feet. I rolled off the roof and fell down on top of the ladder that had fallen into the dog pen, much to the consternation of the dog. I was cut to smithereens. I had all kind of cuts on my body. The dog was mad about the ladder, but now he's concerned about me. When I took an inventory, nothing seemed to be broken, but there was blood everywhere. I crawled out of the dog pen, got into the garage, took a deep breath and realized that I'm bleeding badly enough that I better go get some medical attention. So I kind of wrapped myself up, got in my car and headed off to the clinic (see Figure 14)."

Figure 14
The Christmas Lights Incident

"About an hour and a half later my wife arrived at home; she didn't know where I was. My car's not there, but the garage door is wide open, the dog's running free in the backyard, there's a ladder in the dog pen that's covered in blood, and there's drops of blood that lead all the way around into the garage. She called my boss to see if he knew where I was at, and he told her I had taken the afternoon off to take down the Christmas decorations. So he got in his car, and she got in her car, and they went to the closest clinic, which is where they found me. It was embarrassing enough as it was, but the next day, when I went into work, I noticed that there was a new safety bulletin out that had been issued to every company facility throughout the whole country. The safety bulletin talked about removing your Christmas lights and the importance of really paying attention to what you're doing. It had a very nice cartoon-type drawing, depicting a safety director falling off the roof."

"The point of my story is that even a seasoned veteran, supposedly somebody who knows better, can become complacent enough to actually rush through a job that they shouldn't be rushing through. So there's some rushing and some frustration, certainly a little bit of fatigue, and a whole lot of complacency leading to a poor decision that resulted in a loss of balance, traction, and grip."

"You know," I said almost laughing, "I've heard this story a few times, and I must say that the part where you demonstrate the landing like a cat on the peak of the garage roof like you're some sort of Cirque du Soleil performer, that part always gets a huge laugh from the crowd. But unfortunately, there is a natural tendency for people when they hear this story to say to themselves, 'Well, you should have known better' or 'You shouldn't have been doing what you were doing in the first place' as if these ideas were injury prevention techniques that nobody had ever heard of before. Do they really think that people would respond, 'Oh, if only I'd been told that when I was a kid by my mother or grandmother, then I never would have gotten hurt again?'"

"I think that one of the main problems with safety is we're so quick to be judgmental about somebody who's in a position where 'he or she should have known better.' Instead we should be asking, 'In order for this to have happened, what must have happened to the level of complacency?' and 'Was the level of complacency exacerbated by the fatigue and the rushing, and the fact that he only had so much daylight?' So instead of saying, 'Well, you shouldn't have been doing that in the first place. After all, you're a corporate safety director. You should have known better!' What we should be saying is, 'Hey, if this could happen to him, then it's probably happening to me, and it's probably happening to everybody.'"

"So, we could teach people to recognize that if you're changing what you normally do, and the only reason you're changing what you normally do is because you're in a rush, or because you're frustrated, or because you're tired or a combination of the three, then you could *extend* the self-triggering technique to the rushing, frustration, fatigue in this situation. Instead of self-triggering on the state—in other words quickly thinking eyes on task, mind on task, line-of-fire, balance, traction, grip—you could extend the self-triggering technique by asking, 'Is what I'm doing worth the risk?' or 'Does this state justify the risk?'"

"I heard so many stories from people where they said, 'If I had just stopped to think about it....' Mike Coulter, one of the consultants who worked for us, said, 'The only reason we were rushing was so we could get back to the break room and finish our card game before the weekend.' That decision cost him his right arm and a kidney. You've got to admit, Gary, he's got to be one of the nicest guys you could ever meet on the planet. It was embarrassing because he could set up a training room faster than you and me, and he only had one hand; it was like the one-armed paper hanger thing. But to lose all that just for a card game! Asking 'Is it worth it?' is ludicrous. Of course it's not worth it. But, did Mike stop and evaluate the risk in terms of, 'Is what I'm rushing for—the card game—is it worth it?' The answer would have been instantly 'No.' So I thought, well, if nothing else, we could at least get people to stop and think for a second, to extend the self-triggering technique and ask, 'Does the state justify the risk?' If it is or it does, then yes, you can do what you want. If we could at least get people to stop and think, wouldn't that help? So we put those concepts into one of the *Extended Application Units.*"

"I also heard many stories from people talking about being complacent enough to trust things to memory. After hearing enough stories, I realized the fallacy here was that people quite often think if something is really important, they won't forget. True, if something was really important you would be less likely to forget, but that didn't mean it was *guaranteed* you wouldn't forget. If you started looking at things like people leaving their kids in a car seat on a hot day—49 children in the United States died of hyperthermia in 2010[9]—just because their parent forgot them in the car, you realize that the problem of being complacent enough to trust things to memory is significant."

"I can't tell you how many plants I've been to where you see a sign as you exit that says, 'Did you remove your lock?' Why? Because maintenance people quite often would forget to take their locks off a lockout device once they'd finished their work. You'd see that sign at almost every plant. So I knew complacency also was causing problems in terms of people being willing to trust things to memory or thinking that they might be able to remember a change to a set procedure. It wasn't nearly as frequent or as common as injuries from rushing, frustration, fatigue and complacency, but if there was enough hazardous energy around, one mistake where you didn't recognize the change could be fatal."

"There's an example of this that I'll never forget. It was an electrical utility company. They'd been doing a particular type

---

[9]Jan Null, CCM, Department of Geosciences, SFSU, "Hyperthermia Deaths of Children in Vehicles," <http://ggweather.com/heat/#stats>, accessed April 9, 2012. NOTE: The outside temperature doesn't even have to be really hot for the consequences to be deadly. "When the outside temperature is 70 degrees Fahrenheit, the temperature inside the car can exceed 120 degrees, even when the windows are partially open" according to the Wikipedia article on "Heat illness" (see <http://en.wikipedia.org/wiki/Heat_illness>, accessed April 9, 2012).

of job for 15 weeks in a row. But on one particular day, they had changed the sequence of the job for some reason. The last thing they've got on tape is the dispatcher saying, 'Hey, it's five o'clock Friday afternoon, and I'm outta here. By the way, don't forget to attach the grounds before you hook it up because it's live again, remember?' The next day they found the lineman hanging in his fall arrest harness—dead. There were six possible scenarios, but only two of the six were probable. Of these two, the most likely was that he just forgot the line was energized and reached for it with his bare hand. Obviously, nobody will ever know for sure. So I knew there were problems in terms of complacency: people being willing to trust their memory, or trusting that they'd be able to remember or recognize a change in the environment or with work practices. So I included that in another *Extended Application Unit*. However, that's not it in terms of the problems that complacency can cause."

"In the last few years stories in the media about people driving and talking on the phone, people texting while driving and people adjusting their GPS units have become commonplace. In prior years, distraction faced by drivers, like eating a burger, rubbernecking to look at an accident in the other lane, an interesting conversation with a passenger, or kids yelling in the backseat, had never captured the media's attention quite like the distraction caused by electronic devices. Again, there was a 'How could you?' or 'How dare you?' type of attitude. People who had themselves texted while driving were looking down their noses at other people who were texting. There was a lot of hypocrisy, in my opinion. States and provinces were changing laws about texting, or thinking about changing laws about the use of handheld devices versus hands-free."

"Looking at all of this, I had a different view than 'how dare you.' I thought, 'You wouldn't do that unless you were complacent enough to do it. When you just learned how to drive and you were worried about parallel parking on your driver's test, you wouldn't be texting. If you're going 100 miles an hour (161 km/h) in your car, you're not texting anybody; you're not checking your voice mail; you're not looking stuff up on the laptop; you're probably not opening up hamburgers, French fries or drinks from a fast food restaurant either—because the risk is preoccupying you. You are not going to be doing any extraneous activities. But if you're complacent enough with the activity because you've driven so many miles at 60 miles an hour (97 km/h), you might be willing to talk to somebody on the cell phone or be willing to text something. But you would only be willing to do this if you were complacent enough to do it.'"

"We wanted to get people to start asking themselves questions about their own level of complacency and their comfort level with doing things that they know will increase the risk of making a mind not on task error. In other words, it's bad enough if you're on a long drive trying to keep focused and not having your mind drift off and wander all over the place. It's another thing altogether, though, if you start doing something that will deliberately take your mind off task. Once again, Gary, I know you've got a great story about this one."

"This is one story I don't mind telling," Gary replied. "And it did change me, so that's also kind of interesting. As we discussed earlier, one of the critical error reduction techniques is watching other people for the state-to-error risk pattern. This technique helps you to fight complacency. I had just finished speaking at a conference in Fargo, North Dakota, and I needed to drive from Fargo to Omaha, Nebraska to speak at another conference the next day. So from Fargo, I pulled onto Interstate 29 to go south. It's not an exciting drive. It's pretty flat; there's not much going on until you get to Sioux Falls, South Dakota. From Sioux Falls, South Dakota to Sioux City, Iowa, there's not much going on either. Then there's absolutely *nothing* going on between Sioux City, Iowa and Omaha, Nebraska. So I'm thinking it's going to be a long, boring drive."

"The interstate speed limit there is 75 miles per hour (121 km/h). I'm behind a woman who had pulled on the interstate in front of me. She's going about 80 miles per hour (129 km/h), so I set my cruise control at 80. (I'll admit that, I'm going five over the posted speed limit.) And away we went. There wasn't much traffic, and I guess I kind of convinced myself that 80 was okay, given that there's not much traffic. Things were going well as I travelled down toward Sioux Falls, South Dakota. After a little while, I noticed that she's weaving back and forth a little bit. She was right in front of me, and I thought that she was probably tired. I call that the 'fatigue weave': You're kind of alert, but you're not real alert; you drift over towards the center line, and then you move back; you drift over towards the shoulder, and you move back. You might not even get onto the rumble strips, but you're having trouble keeping the car right down the middle of the road. That's usually a sign of fatigue."

"After a while she started doing something that isn't necessarily a sign of fatigue; it's more a sign of alcohol or drugs. She didn't have her cruise control on, and she was varying her speed significantly. She got way out in front of me, and then she started to slow down. I've got my cruise control set at 80, so I'm getting closer and closer to her. When I moved into the left lane to pass and she realized that she was down to about 60 miles per hour (97 km/h), she sped up real quickly. So I pulled back into the right-hand lane behind her. This little game went on for quite a while. Then finally, she slowed down a lot. I'm tired of this little game by now, so I'm going to pass her. As I pulled out around her, I was watching her. Since she had been weaving back and forth, and driving fairly erratically, I knew SafeStart well enough to know I needed to stay out of her line-of-fire. So I was looking at her and looking at the road and being pretty cautious about going around her. This time she maintained a slower speed, and when I got up next to her, I was in shock again. She was not driving erratically because she was tired; she was driving erratically because she was reading a book. It's not just a little book; it's a hardcover book; I could see it there through the window. I didn't know what page she was on, but she was about three-quarters of the way through the book. She was perfectly happy to drive between 60 and 80 miles an hour reading a book on Interstate 29 headed south towards Sioux Falls, South Dakota. She looked at me, our eyes met and she smiled. There was a bit of shock in her eyes too. I'm kind of enjoying it because she's a fairly young woman. Older men don't get looked at by young woman very often. Then she did something that's kind of unique. She showed me the book, the cover of the book. I'll never forget it—it was *The Da Vinci Code*. Well, I was appalled. Obviously this was the height of complacency. Then she gave me a thumbs-up, as if to say, 'It's a *really* good book.'"

"It is a really good book," I said.

"Well, *she* really thought so," Gary continued. "As I drove toward Omaha, I began thinking about the *Extended Application Unit* on complacency you were working on at the time, and I thought that this would be a great story to put in that unit. So I'm all excited to write the story down, but I didn't want to stop driving until I got to Omaha, Nebraska. I remember turning off of Interstate 29 onto 80 and getting off (I think it was South 72th Street) at the Holiday Inn. I was excited, so I ran into the Holiday Inn, got out my big Franklin planner, and opened it up to a blank page. I started to write down this story trying to make it condensed enough for the workbook, but with enough detail so it made sense. While I was writing I had taken my rental car slip and my map out of my breast pocket. As I continued to write the story out, I glanced up at the map and paused for a moment. Then I picked the map up, and I looked at it. All of a sudden it dawned on me that I have a habit that takes my mind and eyes off the task, just like the woman I'd seen reading while driving."

"Whenever I land at an airport, I get my luggage (I'm glad it got there at the same time I did); I throw it in the back of the rental car; I hop in; I get on the highway; inevitably, I pull the map out and follow it. Now I don't look at it constantly, but I'm looking to see what my next turn is. I'll even put my thumb on where I'm at, and I'll turn the map to the direction I'm going whether it's north, south, east or west. When I thought of what I normally did, it dawned on me that to a certain extent, I'm doing exactly what she did."

"Watching her for the state-to-error pattern made me realize that I had an issue too, even though I had been doing SafeStart training for quite a few years. So the next day I went out and bought a GPS. I know a GPS can take your attention away also, but it's a lot better than looking at the map. In fact, I turned the GPS to the female voice because I've got six children—one boy, five girls—and my wife, so I'm really used to listening to women tell me which way to go. But, the point is that watching her for a state-to-error risk pattern made me realize that I had a gap in my personal safety system, and that I had become so complacent that I was taking my eyes and mind off the road, too."

As I listened to this story, it made me think there's a tendency—I think it's just human nature—to look at people who would be willing to do something in excess of what you'd be willing to do from a risk perspective, and think, "Anybody who drives a lot faster than me on the highway is a maniac, and anybody who drives a lot slower than me is a wimp. If I thought I was driving too fast, I'd just slow down." So what happens is that we all find our own level of safety judgment and risk that we're comfortable with. At the same time there is a general unwillingness on the part of most people to think, "Well, wouldn't other people just be doing the same thing I did, but they just picked a different level?" Instead, most people tend to think that their level is right.

"But I was trying to look at the complacency issue from a different perspective. Instead of trying to say how much complacency is too much in absolute terms, I was trying to get people to think about complacency and the level of their complacency from a very, very personal perspective. In other words, is my level of complacency too high for the risk that I'm about to take? It isn't about whether the handheld cell phone is legal in this state but not legal in that state. It's about whether the handheld cell phone is too much of a risk for *my level of complacency* and my overall skill at what I'm doing? That, right there, is the key to the problem. I was trying to get people to look at complacency—not just because it leads to mind not on task—but to also look at how complacency can affect your decision-making. In addition, I also wanted to get people to see how rushing, frustration and fatigue can affect your decision-making, and whether or not a person could extend some of these critical error reduction techniques to those deliberate risks."

"I also wanted to look at the whole concept of redundancy as it relates to complacency. The concept of redundancy in safety is pervasive. In other words, we couldn't always stop the person from losing their balance, traction or grip, but if we got the person to at least put on a fall arrest harness and attached it to a lifeline secured to an anchor point, you could keep them—in most cases—from falling to their death. By putting seat belts on people, you wouldn't stop the car crashing in the first place, but the seat belts did a really good job of stopping people from going through the windshield and breaking their necks."

"Some safety devices, procedures and protocols when associated with critical safety systems are viewed as absolute necessities, even though they are essentially redundant. Take the example of the cooling system of a nuclear power plant or the primary flight computer of a commercial aircraft. You might have as many as three redundant systems for certain types of equipment because if the first one fails, you need to make sure the second one kicks in. And if that fails, you need to make sure the third one operates. Without such redundancy, the result could be catastrophic. For critical safety systems like these, it doesn't make sense not to have them. However, there are other safety devices, procedures and protocols that some people might incorrectly view as optional, especially when the risk of the activity is perceived to be low."

"Consider a life jacket: If you don't fall into the water, the life jacket is certainly redundant. Unless it makes you look cool or it keeps you warm, you don't need it. You would be less inclined to wear a life jacket if you're a really good swimmer, than if you weren't. But if someone couldn't swim at all—even if the lake was just a mile long—they're probably going to put a life jacket on before paddling a canoe across the lake. Somebody who's an Olympic swimmer may not be so inclined to put on a life jacket because they know they can swim to the shore. But if you hit your head on the canoe hard enough to knock yourself out, it won't matter whether you're an Olympic swimmer or you can't swim at all, you're still going to perish without the life jacket on. But if you didn't fall in, the life jacket would be redundant."

"Locking things out is redundant too, provided nobody starts the equipment up again. The handrail on the stairs is redundant. You don't need to hold the handrail as long as you don't lose your balance, traction, or grip on the stairway. But if you did lose your balance, having that third point of contact and being able to hold on to something is going to keep you from falling down, or certainly limit the likelihood that you'll fall down and go for a wicked tumble on the stairs. So why don't people use the redundant devices, procedures and protocols? It's because they're complacent enough to start with. But the problem only gets worse when you add rushing, frustration and fatigue to the equation."

"So Gary," I said. "The long answer to your short question about 'was there more' is yes: You can extend these concepts and these techniques to various aspects of decision-making without just giving people a lecture on 'Don't do this' or 'Don't do that' or 'You shouldn't have been doing this in the first place.' So yes, we have added several *Extended Application Units* that extend the basic concepts where applicable to deliberate risk, focusing on the different facets of complacency and what it can do to your decision-making, as well as what rushing, frustration, and fatigue can do to your decision-making."

If you look at the state-to-error risk pattern (see Figure 15), you'll get a good summary of Part 2 of this book. The state-to-error risk pattern illustrates how people get hurt over 95 percent of the time. In fact, if you add minimal personal injuries such as cuts, bruises and scrapes, then it can be 98 or 99 percent of the time. So once you understand the fact that this is what's causing so many of our acute injuries, you'll need to totally rethink the way you approach injury prevention. It also necessitates thinking about changing the whole safety management system. It's not just reengineering the physical part of the system; it's going to require us to take a look at the way safety is managed, and this is going to take a lot of resources. It is a whole new way of thinking.

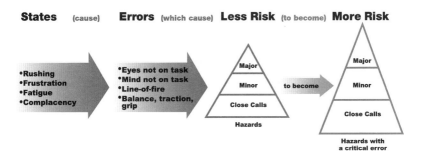

Figure 15
State-to-Error Risk Pattern

If you're still questioning this, it's probably a good time to just pause and ask yourself (as the reader), "Can you personally think of a time you have been hurt, on or off the job—not including sports—where you weren't rushing; you weren't frustrated; you weren't overly tired; you hadn't become so complacent with the hazards that you just weren't thinking about them at that instant; you had your eyes on task and your mind on task; you were looking at what you were doing; you were thinking about what you were doing; you were aware of the line-of-fire;

and you were at least conscious of losing your balance, traction or grip? Can you think of a time you've been hurt when one of those four states (see Figure 16) did not cause or contribute to one of those four errors in the 'self' area? In other words, it wasn't the equipment, or it wasn't the car you were driving that did something unexpectedly; it wasn't the other guy who did something unexpectedly; it was in the 'self' area. Did you ever get hurt when one or more of those four states didn't cause or contribute to one or more of those four errors? That's the question I asked the 20,000 people with the $20 in my hand.

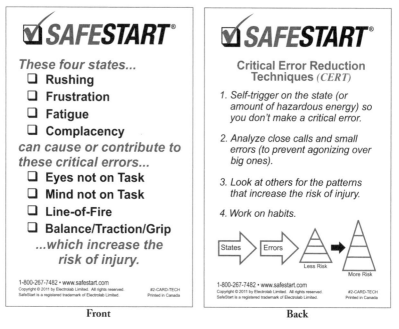

Front   Back

Figure 16
The SafeStart Card

Chances are you probably can't think of a time in your life when one of those four states didn't cause or contribute to one or more of those four errors. So, from an inside-out perspective, the first thing you need to do is to think about it from a personal perspective. Once you've done that part, the next step does require a fair bit of effort and resources (time and money) because now you've got to rethink what you're doing in terms of safety management.

I had an interesting conversation with the manager of the largest site for a big chemical company. He came up to me after the session and said, "You have totally turned my perspective and paradigms on safety management inside out. You have flipped it 180 degrees. For thirty years, I've been asking myself, 'What can I do to keep them from getting hurt?' And now I realize that I can't keep them from getting hurt—*only they can*—which means that I have to rethink, totally rethink, the way I've been managing safety."

PART 3

"Inside Out"—

Reengineering the Workplace

and Community for Safety

CHAPTER 1—*The First Challenge: Yourself*

"I realize there's a certain amount of time everybody needs to be able to convince themselves that SafeStart is the real deal," I said. "That includes even me. I mean, let's face it, I asked 20,000 people with $10 or $20 in my hand. I knew the answer wasn't going to change after the first 5,000. Actually, I knew it wasn't going to change much after the first thousand. The reason I offered people the money was so that when I was talking to people, like the company we eventually partnered with to make the videos, I would have something I could say: 'Hey, I've asked 20,000 people about the four states and four errors with $20 in my hand, and they couldn't get the $20 from me.' Because otherwise, all I could say was, 'Does this make sense to you? Does it explain your injuries?' I didn't have anything at the beginning in terms of data. The only real controlled experiment we had, where the only thing that was added were the SafeStart concepts, was your facility in Iowa. At all the other places I had done this, it was in conjunction with implementing a behavior-based safety process, which would eventually reduce injuries, but not as quickly as SafeStart would. So it took me almost a year to have enough confidence to go forward with the concepts and techniques because, as you said earlier, they really did go against the grain. It certainly went against the party line that was very popular at the time. I think probably the last 10,000 people I asked with the $20 in my hand was just mostly me getting up my nerve."

"So, Gary, when you finally understood all of this—the state-to-error pattern: four states (rushing, frustration, fatigue and complacency) that cause or contribute to four critical errors (eyes not on task, mind not on task, moving into or being in the line-of-fire, or somehow losing your balance, traction or grip)—when you finally realized that this actually was the pattern involved in 97 to 99 percent of all acute injuries on and off the job, once you quit fighting it or looking for holes in it, what did you do?"

Gary paused for a second or two, looked out at the lake and then said, "Well, that's quite a question." He paused again. "You know, it wasn't an easy change for me. I'm an engineer. I'm a safety professional. I've grown up in the compliance world, and there's a lot of security in that world. It's pretty easy: either you are compliant, or you're not. While you don't want accidents to happen, moving from the security blanket of compliance to worrying about human behavior was tough for me at first. Frankly, it was a bit frightening. But I remember seeing a quote by you a long, long time ago: 'When you've reached a point where analysis of your injury, audit and inspection data...' (that's the stuff that we engineers do all the time) '...when that data no longer discovers meaningful trends,...' (in other words, we get no fingers pointing to a certain issue at all) '...it's time to include human factors as part of your research.' We—the safety professionals—did not include those states and errors as part of our research and training. We looked at the analysis of our injury data and, in the early years, that pointed right where we had trouble: If 60 percent of the injuries were in the warehouse, well, guess where you're going to start? You're going to go to the warehouse and start looking at the systems that are in place, and what's actually happening. Then you could take the appropriate actions. But eventually, when you've taken care of most of the compliance issues, what happens is that you don't have any trends, and if you're an engineer, a numbers kind of guy like I am, then you don't really know where to direct your attention."

"When I finally grasped this whole state-to-error piece was when I started seeing the four states and the four critical errors in my own injuries and close calls. And that just started a long journey. It wasn't easy. But when you finally realize that if you're really going to think in terms of injury reductions—how to reduce injuries whether it's at home or at work or on the highway— you understand that you're going to have to make a big change. People were talking about zero injuries, and it seemed an almost impossible task. I'd like to get to zero; I want to get to the point where no one gets hurt, whether it's a family member or a co-worker, whoever it might be. Part of the reason I have not retired is that I think these concepts and techniques are important enough that we needed to take them to the world. That was just over ten years ago. So once I understood it and believed it—I realized I needed to do something."

"As I thought about the state-to-error risk pattern and the implications it had on injury reduction, I thought, 'Wow! This is a big, big change. This is a paradigm shift like nothing I've ever witnessed or experienced. We'll have to redesign, retrofit and integrate human factors into our system, every aspect, every part of our system. That's not going to be easy. I'm going to need a lot of help.' That realization posed a problem because you don't really have anyone else who understands the situation like you do. So, you've got to go through the process of getting other people, eventually everybody, on board. Just like any time there's a big change, there's a number of steps you go through. The whole thing starts when you recognize, 'Wow, the playing field has changed.' Then you check out the rules and the playing field. You learn a little bit more about what's required, and you make an adjustment. That adjustment eventually becomes the new norm."

"We went through something similar in North America not too many years ago when we all of a sudden woke up and realized we had a regulator with safety enforcement teeth, looking over our shoulder. In the United States, it was the Occupational Safety and Health Administration (OSHA). We resisted for a while. It took us some time to figure out what the playing field was like and what the rules were. We still fought it all the way—some people still fight it—but we began to change. Eventually, we adapted to the rules, at least most of us did. Now we have become at least used to working with a regulator."

"Although the concepts and techniques we've been talking about are very effective in reducing injuries, they are much more difficult to get companies to use because we don't have a law or regulation that tells us that we have to do it. In the absence of a regulation, this paradigm shift requires exceptional leadership. We have to get enough people on board in the leadership group to make sure that we've got the support we're going to need to get things headed in the right direction."

As Gary talked, I realized what he was up against back then. He was going to need help. And in this case, it had to be the "willing" versus the "begrudging" kind. When you've got a standard that is developed by the regulator and the regulator has the ability to enforce the standard, you don't have to be a motivational speaker or the best salesman in the world to get people to understand that they need to make the changes the regulator is requiring. But when you're trying to make a huge change that the regulator isn't requiring, it can't be done in the safety office alone.

"So," Gary continued, "like I was saying, we're going to need help in a lot of areas, from management, from engineering and from the employees. These are all really critical areas. We'll need help in introducing the process, help in getting it started, help in the training, help in the different way of thinking about things with respect to inside out rather than outside in. Then, once we've got it off the ground, we're going to need help keeping the process alive and moving forward. It's going to have to have support—forever."

"When you step back and look at what it will take to get everybody to treat rushing, frustration, fatigue and complacency with the same respect they treat physical hazards, you realize just how big an undertaking this is. We'll eventually have to put this into every element of our safety management system, and that's going to take a lot of work. We're also going to need to put it in everything we do, redesigning the systems from the inside out, rethinking the systems and changing our paradigms completely. Plus, it's going to take a lot of work to keep the process alive once we get going. This is a voluntary thing. It isn't like we have a regulator telling us what we have to do. This is something that we have to step through and do on our own. Sure, it's going to require a lot of work. But what excited me and motivated me to do it was that I knew this change could really make a difference!"

"You mentioned there were a number of steps to managing this kind of change," I said. "What were the steps you took? I mean, you managed this change process at three different Fortune 500 companies before you came over and started working full time with us. So maybe you could just walk us through what some of the steps are and what order you took them in."

"The first step was *very* personal," Gary replied. "I had to convince myself that this change was worth the effort. I did that by sitting down and saying, 'Okay, the security blanket I've had all these years—compliance—is not going to go away. We're not taking the guards off the machines. We're not changing anything we're doing to meet the regulations, but we *are* going to look at things a little bit differently.' So I started breaking the task into pieces. It's kind of like eating an elephant. It's impossible to eat an elephant in one sitting; so what you're going to do is eat it one piece at a time. The second step was introducing the leadership group to the ideas and concepts. However, as you can imagine, they, too, had to be convinced that this was a worthwhile project, that it had some merit. You can't really accomplish much without the leadership group on your side. To get them on board required quite a bit of convincing, because they had never heard of anything like this."

"Because this isn't a forced thing that a regulator is making you do," I said, "you're probably going to get resistance from leadership groups in terms of 'Well, we don't *have* to do this.' It's not like we've got a corporate lawyer who's saying the liability risk is huge, so it has to be done. This kind of support requires people believing that they could use these tools to improve safety performance, and maybe even get to zero. But along with that, you also have to take another step and realize that we're trying to get people to improve their own safety-related habits, like moving their eyes first before they move their body, and looking and thinking for things that could cause them to lose their balance, traction or grip. These aren't things that you can easily enforce as a supervisor, either. Now the fact that you can't force people to keep their balance or to look first before they move doesn't mean that it's hopeless. It just means you can't force it. You can definitely influence it, but influencing people requires real leadership. It requires that people follow willingly. One of the things I've noticed is that there are a great number of managers and supervisors who struggle just with managing, directing, controlling and scheduling. When you look at what skill sets they have developed for leading, you realize that they don't have a lot of those skill sets; they haven't had a lot of that kind of training."

"Right," Gary said. "So it's not an easy process because you have to get the entire management team to start with themselves and realize that they need to improve. But before you can do that, you have to realize that maybe *you* need to change as well. One project that I was given really drove that point home for me. Although my work mainly consisted of safety and environmental engineering, this project involved preparing a building for sale. The plan was to sell off a part of our property that contained a building that hadn't been used very much. We were going to use our own maintenance people to clean and spruce it up a bit. The work also included moving a fence so it separated our property from the unused building. I looked the job over with our maintenance people, and we started talking about how we were going to clean up the building. We decided we needed to do some sandblasting to clean the ceiling. Then we were going to paint the ceiling with an off-white color to brighten up the place. We also talked a little bit about the safety issues involved in the project and what kind of equipment we were going to use to get it done. When we'd finished discussing these issues, I left."

"In my mind I thought I gave good instructions, and that we agreed on the equipment to use. I thought that if the work went pretty smoothly, it would probably take about a week for the two maintenance guys to get it done since it was a pretty good-sized building. They'd have to blow everything off, sandblast some of it, and then paint it. Well, two days later they called me up and said, 'Hey, we got it done. Do you want to come out and take a look?' I didn't know what to expect on my way out there. I was a little apprehensive because it only took two days rather than a week, and I worried that it probably wouldn't look very good. But when I got there, it was beautiful, I mean, really well done. I thanked them for a great job and I said, 'Okay, now what are we going to do with the walls? How are we going to blend it in with the ceiling?' So, we kind of planned that out. I walked back up to the office and I got involved with other things."

"But after a while, I got to thinking, 'Wait a second, I had planned on this part being a week-long project and if necessary, I was prepared to see it go even beyond that. How did they do it so quickly?' So I walked back out there to talk to them and I asked, 'How did you do it so quickly?' I mean, it takes time to put up scaffolding. We're talking about a building that had about a 30-foot (9-meter) ceiling, so erecting and moving the scaffolding would have been quite involved. They said, 'Oh, we didn't use scaffolding.' The method they used was quick, but *very* risky. They took two 40-foot extension ladders and put one of them up against a very small channel iron piece in the ceiling. They took the other 40-foot (12-meter) ladder and leaned it up against that ladder. Then they backed a forklift up to the bottom of each ladder to keep them in place, climbed up to the ceiling, tied the two ladders together, and then tied them off. They basically did everything working off of 40-foot ladders without any fall protection, a high-risk activity for sure."

"When I thought about the situation, it dawned on me that I hadn't been enough of a leader; I was just trying to manage the process. I wanted it done right away, and they delivered exactly what I wanted. They knew that I wanted to get the job done quickly, and they knew we had more work to do after that, so they actually just did what I asked them to do. Now, was it risky? Absolutely. Was it the wrong thing to do? Absolutely. Could I have yelled and screamed? Yeah, I could have. But instead of doing that, I apologized to them for not giving them clear enough direction. And that's when I realized that some people are motivated to get things done quickly on their own. That meant we have to be better as leaders to make sure that we completely spell out what we want them to do. You want to be able to leave them working and know that everything they do from there on in will be done properly and in the safest way that it could be done. So, were they rushing? Yes, they were trying to get the job done quickly. It's not a bad thing to try and get the job done as quickly as possible. But they were rushing. I don't know that they were frustrated, although setting up scaffolding certainly slows down the process. So the rushing led to them increasing the risk of the job they were doing. If I hadn't gone back, my feeling is that I would have implied that what they did was just fine. My guess is that if they had a similar job in the future, that's exactly the way they would do it."

I looked at Gary and said, "So I guess the first step is personal? In other words, you've got to realize that *you* have to start making changes yourself."

"Right," Gary said. "People who knew me during the first part of my career have told me that I'm doing things differently now. They say the way I talk and the way I approach things is a lot more comprehensive, and that I take a lot more into consideration than I used to."

CHAPTER 2—*Leadership Support and Influence*

Gary's story reminded me of a manager—a vice president, actually—talking to his senior superintendents about how he had used these concepts and the critical error reduction techniques himself.

"I was at a large hydroelectric utility," I told Gary, "that had implemented the SafeStart process across 1,300 linemen and had achieved a 36 percent decrease in recordable injuries going through the first five units of the course. The vice president of this group heard me speak at a conference in the interior of British Columbia and decided he wanted everybody to hear me present these concepts in conjunction with the first *Extended Application Unit*. He wanted to have me do the training for the whole province. This was a huge job. Up north in some of these places there isn't even a power grid; it's just diesel electric because there are so few people. As I began doing these sessions, it became obvious to me that while they might have seen a 36 percent decrease in injuries, not all of the employees were happy campers."

"The first five units were delivered by the safety department who had never been up a pole or worked with 13,800 volts (or 'Godzilla,' as they said), so there was a credibility gap between the instructors and the linemen. The linemen thought the instructors really didn't know much about their work, or certainly not enough to be lecturing them and insinuating that they made too many mistakes. As you can imagine, it really didn't go over well at all. I remember the manager in one division out on Vancouver Island who told me, 'My job is to introduce you.' Then he looked at me and said, 'Do you have any last words?' I said, 'What do you mean?' He said, 'Well, I'm telling you, it was a nightmare. This was the worst thing that corporate has ever asked us to do, putting everybody through all of this training.' I found out they had brought back a retired safety person to conduct the training for the people in this division. He was 70 years old. His nickname when he used to work there was 'Crash' because of all the company vehicles he had totaled during his career."

"No wonder it crashed," Gary chimed in. "No...seriously, I can certainly see how there'd be a big credibility gap, especially between the young folks who are in their twenties and somebody who's 70."

"In addition to the credibility gap problem," I continued, "it became obvious to me that none of the supervisors and managers had been to any of the training sessions. So when I got to talk to the executive VP of this group, I said, 'Your supervisors and managers didn't go to the training, and they don't understand these concepts and techniques well enough to use them themselves, let alone coach their employees if one of their employees didn't understand them.' He said, 'But they *did* go to the training sessions. I made them go, they all had to go.' I said, 'Well, maybe you directed them to go, but they didn't go, and I can prove it to you. Bring them into a room, and I'll ask them what the four states, the four errors and the four critical error reduction techniques are. I'm willing to bet you my daily fee that they will not be able to remember those four critical error reduction techniques.' He said, 'You're on.'"

"We scheduled three half-day sessions with about 135–150 people attending each session. Not one of the supervisors and managers could remember all four critical error reduction techniques. He basically read them the riot act after that: 'You were required to go, but it's obvious that you didn't go. You will be at these *Extended Application Unit* sessions—you will be at these sessions, or you will not be working here anymore.' He really gave it to them. He then gathered his senior group of superintendents into a room, and I trained them for half a day. They all attended the *Extended Application Unit* sessions, as well. But in my opinion, what really tipped it for everybody was when he spoke to the group about his experience with these concepts and techniques driving to work in rush-hour traffic. I don't know if I've already established that this guy has got a bit of a temper."

"Yeah, I think you did," Gary interjected. "Didn't you tell me that he was Italian?"

I was really surprised that Gary was paying attention since he had been intently watching the guy in the boat fishing for the last five minutes.

"I don't know if that has anything to do with him being Italian or not," I said. "Anyway, he told them he was cut off on the freeway by somebody and had to brake really hard. He said that what he normally would have done was to speed up, pull up alongside the other car, honk the horn, give the driver a dirty look and then maybe cut him off. But instead, he just self-triggered. 'Hey, it's not worth getting hurt over,' he thought, so he backed off, let the guy in, and just kept on driving."

"Having used the techniques for himself, he could convincingly talk to his senior superintendents about his experience as he related the story to them. That had much more influence with his group than his reading them the riot act. I mean, he'd told them to go, and they ignored him. But once they heard him talking about this from his own experience, they really started to become engaged. And once the leadership group was engaged, it made the workforce more willing to put some effort into this. It was so much easier. To make a long story short, they got an additional 36 percent decrease through these *Extended Application Units*, but there was much more going on than just a bit of additional training that extended the concepts. The important thing was the fact that now the leadership group was engaged, where they weren't before."

"Was that what motivated you," Gary asked, "to start producing some material for the leadership group that would help them to understand the difference between influence and control?"

"Yes," I said. "I was also trying to make them aware of the size and scope of the problem we were trying to fix."

CHAPTER 3—*Size and Scope: Quantifying Unintentional Risk and Influencing Change*

Gary looked at me quizzically. "What do you mean by 'the size and scope of the problem we were trying to fix'?"

"Most managers and supervisors don't realize how difficult injury prevention really is," I said, "because they've never thought about how to quantify unintentional risk. When I talk with a leadership group, I take them through an exercise using a risk pyramid. I ask them, 'How many times a day do people in this facility turn or move without looking first?' Then I ask them, 'Do you think this happens less than ten times a day, approximately ten times a day, or more than ten times a day?' The usual answer they give is more than ten times a day per person."

"Then I tell them, 'Okay, let's just use ten because the math is easy. So if we've got a plant with 1,000 people who turn or move approximately ten times without looking first, that means there are 10,000 potential incidents where a person could bump into somebody, bump into something or, worse, move into the path of a piece of mobile equipment like a fork truck, where the injuries could be very, very serious. So that's 10,000 potential incidents a day. Now if we look at the situation over 3 or 4 years, you've got 1,000 days depending on whether the plant works shifts or not. That gives us 10 million potential incidents over 3 to 4 years.' So I write 10 million down at the bottom of the risk pyramid (see Figure 17)."

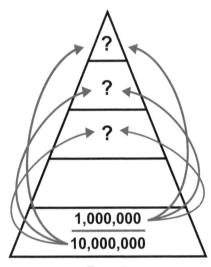

Figure 17
Quantifying Unintentional Risk

"Then I ask the leadership group, 'What are the chances if you've got 10 million down at the bottom, you're going to have zero recordable injuries and zero lost-time injuries?' Well, even accountants who don't understand very much about safety management systems can understand that we're not going to get to zero injuries with that much unintentional risk at the bottom of the pyramid."

"Then I ask the group, 'Okay, how many times a day do you think people momentarily lose their balance—they don't slip and fall; they don't even always have to stick a hand out to grab something like a handrail—approximately how many times a day do you think you momentarily lose your balance?' Well, I've even been trying to keep track of this on a personal basis, and it's certainly more than once a day for me, but it isn't usually more than ten. When I'd ask the leadership group, they'd usually say at least once or twice a day, but not more than ten."

"Now I tell them, 'We'll use one because it's easy math, but I think we all agree it's probably more than once a day; it's just that rarely does anybody bother keeping track of an incident like this. Suppose, for example, you catch your foot and lose your balance. Very few people would fill out a near miss report at that stage of the game. So while the number of times people lose their balance is probably higher, we'll just use one for easy math. If we take the same plant of a 1,000 people, in 3 or 4 years, that's 1 million potential slips and falls. What do you think the chances are that none of those are going to be recordables? How much of your own money would you be willing to bet that with 1 million potential slips and falls, none of them are going to end up as recordable injuries or as lost-time injuries?'"

"Managing all this unintentional risk is also difficult," I continued, "because managers and supervisors believe that they can manage it the same way they manage other things. But can a manager force somebody to keep their balance? Can a manager force somebody to move their eyes first before they move? The answer is 'No.' You can't force them, but you can influence them. Probably the best way you can influence them is just by referring to your own experience by saying, 'This is working for me; this is working for my family. I'm seeing results; I hope you see some too.' When explaining to the leadership group that they need to pull people to where they are, I'd say, 'You need to show them that where you are now is a better place than where you were, and *invite* them to come over to where you are. You can't force them to take these concepts home to their family any more than you can force them to drive safely once they leave the plant. But you can influence it. And you can influence it with 'Look how well this is working for me. Wouldn't you and your family like this, too?' I've also used a piece of rope to represent influence and a stick to represent control. When talking about the difference between the two, I'd say, 'Influence is like a piece of rope: you can't push, you can only pull.'"

"Moving from control to influence is difficult because the traditional safety management system is built on compliance. The regulation requires specific actions; the regulator ensures that the company is complying with the regulations; the company ensures its employees comply with the written program based on the regulation. It's all about tight control from the highest level to the lowest one. If safety were merely compliance, control might be the only thing you need. But safety is more than compliance; it's all about injury prevention, and that requires a different method to achieve results. The reason I'm mentioning this is because some companies have gone too far with the hammer and the stick approach. Some have even gotten to the point of out-and-out intimidation."

"I remember hearing a story about a refinery down near the Gulf where the plant manager stood up in front of 1,000 people and said, 'You will not have a recordable injury here, and if you do, I promise you, you will not have two (because you will not be working here).' His little speech didn't give anybody much incentive to report injuries. If a manager or a supervisor has done a lot of that in the past, the row they have to hoe now is going to be really long. I mean, it's a long enough row to hoe as it is, but now they're going to have to somehow backtrack on using control as their primary method in order to be able to get to a position where they can have some influence. I think anybody who has kids, especially teenagers, knows that if you push on the control button too often, the influence button doesn't work anymore."

Gary looked at me and said, "You're right, those that have been using command and control are really going to struggle with this. If you want to see somebody struggle and generate a lot of anxiety, then give them a job that they're not skilled enough to perform. We're going to have to help them, it's going to take a while, and it's going to be difficult to get them to really grasp the necessity of using influence."

We both felt that there would be some individuals in every company who would really struggle to move from control to influence. However, I also knew that, to a certain extent, everybody has a personal struggle with these skills and techniques. Just because I developed them doesn't mean that I never turn without looking first. It doesn't mean that in the past ten years I have never stubbed my toe. Do I stub my toe as often as I used to? No. But am I perfect? Of course not.

The easiest way for a manager or supervisor to show their employees that they have changed is to talk to them about the skills and techniques from their own personal experience. "Here's what I'm doing; here's what I'm working on. Here's what my family's working on; here's how it's helping me. And I hope it helps you." But all this assumes that you're trying to improve your own safety-related habits, working on being able to use the critical error reduction techniques in real time, and taking these concepts and techniques home to your own family. Getting better at the skills and techniques takes practice; it's like a skill set. And just like a sport, everybody knows you can't quit practicing if you want to stay good at it. Sure, even if you practice, you'll have little slip-ups. Even the best golfers hit the ball into the water and into the woods—not as often as I hit it into the water or the woods—but every now and again, they do. So all you really need to do is just start talking about your own personal experience with the habits, skills and techniques, and you'll be able to influence people. This strategy has even worked for some people who have pushed the control button too often or who have been a little too heavy-handed with the stick.

The importance of the leadership group in the success of the process has been demonstrated time and time again. Without their support the initial training wouldn't even get off the ground. They're the ones who help schedule the training sessions and get themselves and their team there. So we need their support to roll out the training. But we've also found that the more they personally got involved in the process, the more they could talk about their experience using the skills and techniques, and the easier the rollout went. One part of the leadership group, the safety department, has the challenge of integrating these concepts and techniques into the safety management system to ensure the ongoing success of the process. It's a daunting job when you realize that integrating the inside-out approach will affect each element of the system.

"I wouldn't imagine your advice to other safety directors would be, 'Let's tackle all 20 or all 25 elements at once,'" I told Gary, "because you talked about eating the elephant one bite at a time. I know that the exact order of some of the steps you need to take are company specific, but I remember you telling me once that probably one of the easiest things to do, and one of the best things to do in terms of reducing injuries and integrating this into the culture, would be working these concepts and techniques into your near miss or near hit reporting system."

CHAPTER 4—*Integrating Near Miss Reporting and Risk Assessment*

"Near miss reporting is one of the most valuable and robust safety processes we have, as long as it is done correctly," Gary said. "Just think about it: We have the opportunity to identify gaps in our safety system, system gaps capable of causing or contributing to very serious injury or property damage. It doesn't matter if the gap is due to the process or its design, equipment or human factors—we find out what it is without having an injury or equipment damage. However, the problem with near miss reporting has always been participation, or to be specific, the lack of participation. This is somewhat understandable since 97 percent or more of the time, the gap in the safety system isn't the equipment or the other person causing an incident. It's us making an error. The error is something we never intended to make, but we made it nonetheless. No one really wants to admit to making any kind of an error or mistake when the result would be a comment like, 'What was he or she thinking? No one in their right mind would have done that.' But lack of intelligence or mental ability really wasn't the problem. They probably just weren't thinking about what they were doing. This culture of blame has led to very low participation in even the best-run near miss reporting systems. The gaps that are being reported are almost always machinery or issues with other people. As a result, the focus moves away from the real problems to what the reporter wants us to hear, which doesn't do us any good. However, when people have been exposed to SafeStart, they now have a common language to describe the near miss and are much more willing to share experiences that could have been serious because they have participated in the discussion groups during the training. They lose their fear of reporting a near miss and actually start looking for gaps in the safety system in real time."

"The interesting thing about SafeStart is that the concepts are so effective that people begin to use them as soon as they are introduced to them. Let me give you an example. I was conducting a very technical accident investigation and root cause analysis training for about 30 seasoned design and production engineers—riveting training, as you might imagine. Now since any root cause training has to include human factors, and given my belief that the SafeStart concepts are the best way to quickly promote understanding of how human factors are an issue that cannot be ignored, I introduced the concepts just before lunch. We had a brief discussion, and just touched the surface of the states and errors."

"During lunch, I realized that I did not have my planner with me, and that the itinerary for my flight to Brazil later that afternoon was in the planner. I thought I must have left it at home. So I called my wife to have her look for it. Well, over the years my wife has grown tired of my 'emergency' requests, and she was less than enthusiastic about the task I had asked her to accomplish. I was sure it was in the back of my other car, and I should have just enough time to get home, pick it up and get to the airport. She called back on my cell phone and said she couldn't find it. After I finished eating, I got up to drop off my dirty plate at the counter while thinking about what I was going to do. When I got there, I bumped the plate into the counter and dropped it (rushing and frustration leading to eyes and mind not on task). The two engineers who saw this sterling event asked me what state I was in and what error I made. We all had a good laugh. Just about that time, I got a text from my wife: 'Found it, you idiot.' What a relief!"

"When the class resumed just a few minutes later, I started the session with a discussion of what had just happened, and the whole group was more than happy to fill in my states and errors. There was lots of laughing, but an interesting thing happened: One of the participants shared a story about himself under a similar circumstance, and then another shared a story, and then another. It was as if we experienced a cultural shift right there in that room! The common language that the SafeStart concepts provide, combined with the opportunity to openly talk about states, errors and risk, had created an incubator of sorts for organizational and cultural change. When an openness to talk about human error and the associated risk spreads throughout a company, then leadership does not have to push near miss reporting nearly as hard—it happens naturally. All they have to do is provide a little organization to make reporting near misses easy and convenient, in order to have a huge amount of accurate information available about the potential risk present in the organization. So now, if you work on removing or mitigating the risk that was reported, then your safety performance will improve dramatically."

I told Gary that I agreed that probably one of the best things to do to integrate these concepts and techniques into the safety management system was to start with near miss reporting. However, most companies struggle with near miss reporting. It is part of their safety management system, and they have the cards, which they've made easy and accessible enough. But they still don't get very many cards turned in. It's rare to get as many reports of close calls as are actually happening out there. Typically, what you get are things that people would not be afraid to report like a wire rope breaking. But the chance of them reporting a close call when they lost their balance and almost fell is very, very low—almost non-existent.

"Well," Gary said, "they might not even recognize it as a near miss because they think that's just part of life."

I told Gary that not even recognizing it as being a near miss in the first place was part of it, but it could also be that they didn't want to report it. Some might even think it wasn't worth reporting. One thing that happens when you do SafeStart training is that people all come to grips with the 'self' area. They realize that it's rarely the equipment or the other guy who initiates the chain reaction that results in an incident. It's almost always a critical error that was caused by one of those four states (rushing, frustration, fatigue and complacency). Everybody comes to realize the 'self' area is well over 95 percent of all accidental acute injuries on and off the job.

When safety professionals have told me about a marked increase in near miss reporting (see Figure 18), I've asked, "Why do you think there was such a dramatic increase in near miss reporting?" One safety director told me, "We're not exactly sure, but my thinking is that as we went through all of this SafeStart training and everybody realized the importance of the 'self' area, how high the 'self' area was and that we were all in the same boat together, so to speak, I think it just freed everybody up. They realized, 'Hey, I can report this. Other people can report this. We can all report this kind of stuff.'" So learning about the importance of the 'self' area will free up your people to report things they previously wouldn't have reported. It will also help them recognize as near misses things that before would never have occurred to them to report.

The safety director also showed me a chart that summarized cumulative injury potential for one month (see Figure 19). If you just looked at the first two

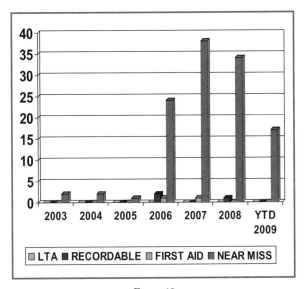

Figure 18
Near Miss Reporting

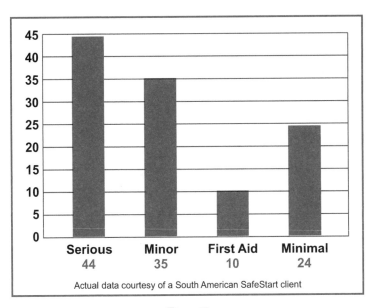

Figure 19
Cumulative Injury Analysis
(September 2009)

179

categories, they had the opportunity to take corrective action with a total of 79 serious and minor potential injuries before they actually became real injuries. So, arguably, they got the chance to prevent 79 recordable injuries. Of course, the statistical likelihood is that not all 79 of them would have ended up as recordable injuries: some might have been near misses or minimal injuries, cuts or bruises, something like that.

We all know that it's rare that somebody gets hurt the very first time they do something that's unsafe. Eventually, though, the unsafe practice leads to an injury. So if we can uncover this unsafe practice, we can start taking corrective action. This action might include some physical redesigns here and there, but it could also include getting the employees to understand which critical error reduction techniques they could have used to prevent this from happening again. That will have a very positive effect on preventing injuries. So I definitely agreed with Gary that one of the first positive steps for reducing injuries would be to integrate the concepts and techniques into the near miss reporting system, because it is going to have a real impact on preventing injuries.

A simple way to capture near miss data would be to incorporate these concepts and techniques on a pocket-sized card (see Figure 20). These cards can be turned in to an employee's supervisor or safety representative or dropped off in boxes conveniently located throughout the facility.

"What other things are important when integrating human factors into a near miss reporting system?" I asked Gary.

"Let me just say that when you fill out a near miss report, use that time to reflect on why it happened, how it could have been worse and how it could have been prevented. If the close call happened

SAFESTART®

Near Miss Report

What Happened?

Most Probable Outcome

- ❑ Fatal
- ❑ Major
- ❑ Minor
- ❑ Minimal

Name (optional)

Dept. *Date*

1-800-267-7482 • www.safestart.com

#2-CARD-NEAR MISS
Printed in Canada

Front

What Happened?

- ❑ Rushing
- ❑ Eyes not on Task
- ❑ Frustration
- ❑ Mind not on Task
- ❑ Fatigue
- ❑ Line-of-Fire
- ❑ Complacency
- ❑ Balance/Traction/Grip

What CERTs Could Have Prevented It?

- ❑ Self-trigger on the state or amount of hazardous energy.
- ❑ Analyze close calls and small errors.
- ❑ Look at others for the patterns that increase the risk of injury.
- ❑ Work on habits.

Corrective Action Required?
(To be completed by Safety Department)

❑ Yes ❑ No

Assigned to: ______

Back

Figure 20
Near Miss Report Card

because you were in a rush, frustrated or overly tired, ask yourself why you weren't able to self-trigger soon enough. If complacency was the primary cause, try to identify the safety-related habit that you need more work on, and work on it. Once the card is turned in and someone in the safety department tabulates the information, make sure you thoroughly review the incident for any corrective action that needs to be taken. Remember, the goal is reducing injuries. So, if maintenance or engineering needs to get involved, get them involved. If an administrative control would have helped to prevent the near miss, implement it. Do whatever is necessary to ensure that the close call never turns into an injury." I smiled. Gary was such a traditionalist.

"Okay, so near miss reporting is probably the first thing you'd want to do," Gary continued. "The next proactive approach is to look at a couple of systems that we're all familiar with. Most organizations use techniques like job safety analysis (JSA) or job hazard analysis (JHA), and some actually perform a risk assessment, something that's a little bit more complicated. But rarely, if ever, do we see any human factors in those JSAs or JHAs, or the risk assessment. They all look at compliance or physical hazards or quantifiable risks. Take the JSA, for example. It's supposed to help us look at the risk associated with every step that a person does on a job, or every step that a group of people do on a job. By going through it step by step, writing down the potential for injury that we have in each one of these steps, and then writing down the corrective action or preventative measures taken for each one of those steps, the idea is that hopefully, by going through this process step by step, we didn't miss a hazard that could get someone hurt. The problem is that human factors aren't considered at all. To further complicate the issue, some jobs tend to create the need, at least a perceived need, to rush, and sometimes the system itself forces, encourages or directs employees to do certain things that increase the risk. These issues are hardly ever included on a JSA."

Once again, I started thinking that some jobs by their very nature have these states built into them. For instance, you get a flat tire on the freeway. It is not something you planned, and it is going to cause delays. There's a certain amount of effort you would prefer not to make—getting the tire out of the trunk and everything else. There's also the expense of either fixing the tire or replacing it. It also occurred to me there are workplace tasks where you could anticipate one or more of the states being present even before the job starts, or occurring during specific steps of the job.

"So what's the best way to integrate the concepts and techniques into a JSA or JHA?" I asked.

"I think the best way is to simply add a fourth column to the JSA (see Figure 21). The first column lists the specific steps to complete the job. The second column is new. It allows you to account for the four states right up front. The third column is

Sequence of Tasks (Steps to Job)	Human Factors (States that Lead to Errors)	Potential Sources of Danger (Hazards & Errors)	Preventative Measures (Physical Barriers & Critical Error Reduction Techniques)
1. Park vehicle.	*Check your states:* ☑ Rushing ☑ Frustration ☐ Fatigue ☐ Complacency	1) *Hazardous Energy:* ☐ Electrical ☐ Chemical ☑ Mechanical (Hydraulic, Gravity, etc.) ☐ Thermal ☐ Other 2) *Critical Errors:* ☑ Eyes not on Task ☑ Mind not on Task ☑ Line-of-Fire ☐ Balance / Traction / Grip *Examples:* a) Can be hit by passing traffic (Line-of-Fire) b) Can be hit by vehicle if it rolls c) Vehicle may roll on uneven, soft ground	*Physical Barriers:* a) Drive to area well clear of traffic. Turn on emergency flashers b) Choose a firm, level area. c) Apply the parking brake, leave transmission in gear or in PARK, place blocks in front and back of the wheel diagonally opposite to the flat. *Critical Error Reduction Techniques (CERT):* ☑ Self-trigger on the state (or amount of hazardous energy) so you don't make a critical error ☐ Analyze close calls and small errors (to prevent agonizing over big ones) ☑ Look at others for the patterns that increase the risk of injury ☑ Work on habits
2. Get spare tire and tool kit.	*Check your states:* ☐ Rushing ☐ Frustration ☑ Fatigue ☑ Complacency	1) *Hazardous Energy:* ☐ Electrical ☐ Chemical ☑ Mechanical (Hydraulic, Gravity, etc.) ☐ Thermal ☐ Other 2) *Critical Errors:* ☐ Eyes not on Task ☑ Mind not on Task ☐ Line-of-Fire ☑ Balance / Traction / Grip *Examples:* a) Lifting spare may cause strain.	*Physical Barriers:* a) Turn spare into upright position in the wheel well. Using your legs and standing as close as possible, lift spare out of vehicle and roll to flat tire. *CERT* ☐ Self-trigger ☐ Analyze close calls ☐ Look at others ☑ Work on habits

Figure 21
JSA with Integrated SafeStart Concepts and Techniques

used to consider the potential hazardous energy and critical errors that an employee might experience when performing the job. The fourth column lists the preventative measures including the critical error reduction techniques used to mitigate the potential hazards of the job."

"That sounds good," I said "but if the safety department integrates the concepts and techniques into the JSAs, how will they know the specific steps that are problematic for one of the states?"

"I'm glad you asked that," Gary said, "because it's important to have the employees who perform the task help redo the JSAs. Most safety professionals know that you probably should have had the employees doing the JSAs all the time. But if you haven't in the past, it's critical at this point because now we want to know what risks are involved with rushing, frustration, fatigue and complacency. Employees who perform that job or task are the only ones who know what really happens. So what we're trying to do is help them do a better job of assessing risk. They're the ones who do it, and they know what's happened in the past. What we need to do is show employees how to use the concepts and techniques when creating or re-evaluating a JSA, so they can start doing risk assessment from the inside out."

I told Gary that I had talked to safety professionals who were trying to develop matrices to help employees recognize hazards and assess risk (see Figure 22). But time and time again, I heard them lament the fact that their employees didn't recognize hazards or do a proper risk assessment. When I looked at what they were trying to use, I saw that the tools they were giving employees were flawed right from the beginning because they relied on the outside-in approach. The risks they were evaluating were all centered on the physical work environment. But they were ignoring the crucial risk, accounting for over 95 percent of all acute injuries, found in the "self" area. Unfortunately, no one evaluates that risk very well or very

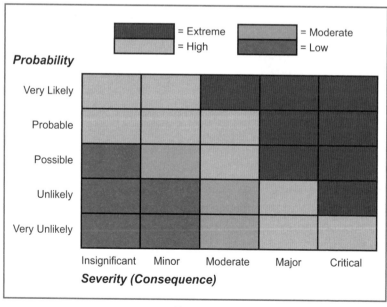

Figure 22
Qualitative Risk Assessment Matrix

readily.

When safety professionals would ask me, "What do you have on hazard recognition?" I would say, "What do you mean?" "Well, my employees can't recognize hazards," they'd respond. "What do you mean, they can't recognize hazards?" I'd ask. They'd say, "I asked the guy why this happened, and he would tell me, 'Well, I didn't see it.'" Then I'd say, "You're trying to tell me that he wouldn't recognize that a nail protruding from a board was a hazard?" They'd say, "No, I...I didn't say that." Then I'd reply, "Well, is it that he didn't *see* it?" They'd say, "Well, yeah," and I'd tell them, "That's different from not being able to recognize a hazard for what it is."

When a four-year-old girl walks into a room and there's a wrought iron coffee table that has a very sharp, 90-degree corner, that four-year-old girl may not recognize the corner on the coffee table as having the same kind of injury potential as an adult might. That kind of hazard recognition needs to be taught. But I think what those safety professionals were talking about is not seeing the hazard in the first place. In other words, the guy turned without looking first, stepped on the nail, and only recognized the hazard when the nail went through his foot. That is not the same as not being able to recognize that a nail protruding through a board was a hazard. Not seeing a hazard or not thinking about a hazard you already know about (eyes and mind not on task) is a much different problem than not being able to recognize the hazard for what it is in the first place.

Here's a personal example that will help demonstrate the difference. Our family was over at my mother-in-law's house for Thanksgiving dinner, you know, the huge turkey spread, when it started snowing. My mother-in-law told us we shouldn't drive home; instead, we should stay there overnight. Looking out the window at how much it was snowing, I said, "Oh, I don't know if it's really *that* bad." Before we left, we checked the Internet and looked at weather reports. We turned on the radio for any news about the highway being closed. We did all kinds of physical risk assessments. But until we were in about the third hour of the drive home and I was so tired I could barely keep my eyes open, no one, including me, ever asked, "What is the risk of driving four hours at night after a big turkey dinner and falling asleep at the wheel?" We didn't think about that risk at all. And yet people falling asleep at the wheel cause approximately 25 percent of the fatal car crashes in Canada and the United States.[1] The percentage of fatal car crashes due to bad weather is about 16 percent.[2] Of course, if you live in a place that doesn't have sleet, snow, fog or ice, the percentage would be lower.

There's a natural tendency for people to look at the physical environment, the physical hazards, when they're doing a risk assessment. Questions like "What is my state of mind?" or "How complacent have I become with this activity?" are rarely asked by people doing risk assessments. Yet complacency is a contributing factor (major or minor) with almost every accidental acute injury. Ask people this question: "Can you think of a time during your life when you've been hurt—other than sports—when you were

---

[1] The percentage cited is our best effort to make sense of a confusing array of data. Unfortunately, the inconsistent crash-reporting procedures and lack of common codes among various jurisdictions allow for the wide range of variability in fatal crash statistics associated with driver fatigue and drowsiness. Here are several statistics: 52 percent (NTSB single-vehicle heavy truck study), 50 percent (study of two busy U.S. highways), 25% (study by German insurance companies), 18 percent (1994 Australian study) and 14% (Canadian university study). See SmartMotorist, "Driver Fatigue Is an Important Cause of Road Crashes," <http://www.smartmotorist.com/traffic-and-safety-guideline/driver-fatigue-is-an-important-cause-of-road-crashes.html>, accessed April 9, 2012.

[2] Ghazan Khan, Xiao Quin and David A. Noyce, "Spatial Analysis of Weather Crash Patterns," <http://www.topslab.wisc.edu/publications/xiao/JTE_weather_safety.pdf>, accessed April 9, 2012.

thinking about what you were doing and the risk of what you were doing at the exact instant when you got hurt?" If you have a hundred people in front of you, you'll rarely have more than one or two hands in the air. Complacency is a significant contributing factor, but not many people take it into account when they're doing risk assessments.

A few years back, I was the opening keynote speaker at a big petroleum conference. Because I didn't have to leave until the next day, I stuck around to hear the closing session. The closing keynote speaker was a very interesting guy—a long-haired surfer dude named Bruce. He actually had a wind surfer strapped to the roof of his 4x4, which I saw later in the parking lot when I talked with him. His real job was working as a whitewater rafting guide on the Yukon River. But as an adventurer, he'd rafted the Blue Nile and the Amazon, traveled across the Sahara Desert on camels, and summited Mount Everest. With one of his close friends, he'd done a number of first descents, skiing down some peaks in the Rocky Mountains. All these adventures he documented with phenomenal video and still pictures.

Much of his presentation was just incredible visuals. We watched a videotape that gave you the feel for what it was like being at the top of Mount Everest or on the Blue Nile River or the Amazon. It was just amazing video. He showed us the people on the various excursions he'd been on: the Sherpas, their families, and the people involved in all of these treks. Then he started talking about the risks that weren't so obvious to all of us watching the video. You can't see that it's almost minus 30 degrees Celsius (-22 degrees Fahrenheit) when you're climbing to the top of Mount Everest. He also talked about the thin air and explained how the low oxygen level was responsible for how slowly everybody was moving. He talked about how the people during the treks were continuously doing risk assessments. They monitored the weather to ensure they were not going to get caught in a storm, which could prove fatal. They inspected and assessed their equipment. The risk was obvious: If the rope broke or the D ring failed, you could fall to your death down the mountain. So they continually monitored the obvious things.

"But rarely did I find people assessing risk in terms of themselves on a personal basis," he said, explaining that they didn't ask questions like, What if I make a mistake? or What if I sprain an ankle and can't walk? "At that elevation, carrying someone is too hard. So a sprained ankle or wrenched knee could be a death sentence," he said. He recounted all of the little mistakes people can make. He told us a story about how one time he nearly forgot the GPS and the satellite radio phone at the camp site because he didn't double-check before they left. He was pretty sure he had it, but he didn't. Those are the little mistakes that complacency can cause.

At the end of his presentation, he again showed photos of some of the people who had accompanied him on various expeditions and their families. Then he showed us their burial site with its marker or tombstone, or, in some cases, a makeshift grave in the snows of Mount Everest. He told us that all of these people died because of the little mistakes that complacency can cause in these unforgiving environments. As he concluded his presentation, he gave us his take on risk assessment: Over 80 percent of the risk in these extreme environments where he had been did not involve the weather, the equipment, or the other people with you on the expedition. The main risk was *you* and the simple mistakes that complacency can cause. "More than 80 percent of the risk involved you," he said, "but rarely, if ever, did people take that into account when they were doing risk assessments." This was a very powerful presentation, especially the way he introduced all of these people and then showed you their tombstones.

I ran into Bruce later in the parking lot, where his car was parked two spots away from mine. We talked about the four states, not just complacency, and the four errors. I briefly explained the four critical error reduction techniques. Then we talked about the three sources of unexpected, and I told him what I had found (at this point in time, I had asked more than 100,000 people) was that the risk in the 'self' area, which actually initiated that chain reaction, was well over 95 percent. I told him for some people, the source of the unexpected event for 97 to 99 percent of their acute injuries had been in the 'self' area. He said, "I don't have any data on this 80 percent. All I know is that it's really high, and I wanted people to realize that it was really high. I just picked 80 percent because I wanted them to realize that when they do risk assessments, they've got to look at themselves." We talked for a few minutes. I told him that his presentation was great. We exchanged business cards and parted company.

Redoing your JSAs and rewriting your procedures so human factors become part of your safety management system will make these concepts and techniques much less likely to fade away or to become a flavor of the month. But there's a secondary purpose for integrating the states, errors and critical error reduction techniques (CERTs) into your JSAs: worker involvement. You need to involve your workers since they're going to know, much better than anybody else, the problems that rushing, frustration, fatigue and complacency can cause while performing their job tasks. Getting them involved starts the wheels moving in terms of how they assess risk in the future. With this change, your employees can now assess risk from the inside out. They can assess their personal risk, evaluating the states they are in and their current level of complacency. Getting them involved also teaches them to be able to monitor their level of rushing or frustration or fatigue as they're working, and to consciously recognize that these states actually increase risk. This is so much more powerful a reason than simply rewriting the procedures so the concepts and techniques don't fade away.

Once you've engaged your employees in integrating the concepts and techniques into JSAs, you'll get improved risk assessment from every individual on the shop floor. If you do that, you'll probably have enough buy-in, enough personal belief in these concepts, from employees, as well as from the management team, so that you can move these concepts into incident investigation.

So I asked Gary, "After we've added human factors to the JSAs and are seeing improved employee risk assessments, would this be a good time to start integrating these concepts and techniques into the accident/incident investigation process?"

CHAPTER 5—*Enhancing Accident/Incident Investigations*

"This is a bit dicey," Gary said, "because if it's true that human error causes over 95 percent of all accidents—and I believe it is—then this could potentially turn into something ugly: a blame game. So you've got to be really careful with the timing of integrating the concepts and techniques into accident/incident investigations. Once an accident occurs, you can't do anything to undo it. After it happens you want to identify all of the root causes so you can be proactive and focus on preventing similar incidents in the future."

"You know, often the action taken to prevent recurrence in these accident/incident investigations is a joke," I said. "The investigator knows they can't say 'Be more careful,' so when they fill out the accident report, they write, 'Told the employee to be more *cognizant of his situational surroundings* in the future,' or some other fancy, esoteric, polysyllabic word for 'be careful.'"

"That's exactly the problem," Gary said. "Most accident investigation processes never really get to all of the root causes because they often ignore the human factors. Most accidents aren't initiated, in terms of the primary cause, because a piece of equipment broke or because somebody else did something that you didn't expect them to do. So when you're starting to talk about the 'self' area, you need to have a little buy-in first. That means you need to make sure you integrate the concepts and techniques into the near miss reporting system and involve employees in the revision of JSAs *before* you start to integrate your accident investigation process. Addressing it after these steps will ensure that you have sufficient buy-in from employees so that what you do will not be misunderstood."

"You also have to be careful with the timing of when you talk to the employee about any human factors that might have been involved. That's where the difference between control and influence becomes critical. If you really want to find out what human factors might have been involved, you probably should wait a few days after the traditional accident investigation to ask. Like I said, if you have the wrong timing, things can get a bit dicey. The bottom line is you've got to make sure that you're finding out what the root causes are. If the triggering event was a critical error, then you have to find out what the state was. Sometimes you'll get a real surprise as to what caused the employee to be in that state when they made an error that started the chain reaction that caused them to get hurt."

I wouldn't want to argue with Gary, but I did wonder if you already had a culture that was quite positive and where employees accepted these concepts and techniques very readily, then you might be able to integrate accident/incident investigation sooner. Without that culture, however, he's right. We've had a great deal of push back from employees when their companies moved on this issue too quickly. Interestingly enough, outside North America, where cultural paradigms of accident causation are a bit different, you find very little push back to these concepts and techniques. Nobody's ever told people in South America that all injuries are management's fault or the result of management's oversight. Nobody's suggested to them that when you get hurt, it's time to sue somebody, like a personal injury lawyer here might do. You don't see helpful signs like 'Know your rights; don't get hurt twice' on park benches, bus stops or prominently displayed on the cover of the Yellow Pages. The mindset is different, so you don't necessarily have a backlash. But what I have seen when safety professionals insisted on introducing these concepts and techniques too early before they had sufficient employee buy-in wasn't pretty. Employee reaction was loud and clear: "Oh, great, now they're just going to blame us for everything. It used to be that at least if you made a mistake, an honest mistake, they would say, 'Hey, we all make mistakes.' But now they're just going to blame us for everything."

I saw this happening at some companies where they were under a lot of pressure from corporate about their safety numbers. The safety professionals decided that if somebody did get hurt, they would talk to them about the states and the errors during the traditional accident/incident investigation. If the employee actually admitted that there was some rushing, frustration, fatigue or too much complacency, and it caused one of these four critical errors, the employee had to repeat the whole SafeStart course again since they must not have learned it well enough the first time. As you can imagine, that did not improve morale. Managers have a tendency or natural urge to control things, especially when corporate is telling them they're responsible and accountable for incident rates. With the pressure they were facing, their commitment to "influence" waned as they reverted to "control."

But I think it's only natural for a lot of safety professionals, since they are so heavily involved in accident/incident investigation, to want to jump the gun and get into it just a bit faster than they should. They need to wait until the overall acceptance level of these concepts and techniques in the workforce is right. Otherwise, they're moving ahead of where their culture is. If their culture hasn't moved to accepting this yet, let alone *embracing* it, and then they throw in something where there is the potential for blame, it could cause a lot of problems. So Gary is correct: If you start with near miss reporting and then you work on your JSAs/JHAs, you certainly limit the risk that you'll have a huge backlash when integrating the concepts and techniques into accident investigation.

However, there is one more thing you can do to prevent a lot of problems: encourage employees to use the DVD *Hurt at Home* with their families. This DVD, part of the *SafeStart Home* series, was designed to help parents and babysitters talk to their children about accidents that occur in the home. The program demonstrates how to talk with children about the states and errors. Of course, you wouldn't do this while their finger was still bleeding or their ankle was turning black and blue. But after you applied the adhesive bandage or ice, after you console them with their favorite book or TV program, after the injury is taken care of and they have calmed down a bit, it's time to apply what they have already learned from the DVD *Taking SafeStart Home*. It's time to talk about the states and errors, and which critical error reduction technique or techniques might help prevent a similar thing from happening in the future. If the child was rushing, you could say, "Hey, did you have your eyes on task?" "No, I didn't." "Well, what do you need to do?" "I guess as soon as I recognize I'm rushing, I should keep my eyes on task, look for line-of-fire, and anything that could affect my balance, traction, grip." "Yes, that's what you need to do in the future." Trust that the pain they experience will help to motivate them to apply the technique next time.

If employees have taken SafeStart home to their families as they were encouraged to do at the end of the five-unit course, introducing the concepts and techniques with *Taking SafeStart Home*, and applying the concepts and techniques after an accident has occurred at home, there should be little resistance when enhancing accident investigation in the workplace. In other words, when you start with the employee and their family, and when they see positive results with their family, there aren't many problems bringing it into the workplace to help the employee and their co-workers. We've found that it significantly minimizes a lot of the resistance. If most of the employees are doing it at home, or at least those with children are, you probably don't have to worry too much about bringing it into the workplace.

The process for including these concepts into accident/incident investigation is fairly straightforward. As Gary suggested, wait a few days—maybe a week at most—after the traditional investigation is done, and then get someone to ask the injured employee what states and what critical errors were involved (see Figures 23 and 24). The person who is asking the questions needs to know how to ask them in a "non-threatening, happens to us all" manner. For instance they might say, "Hey, we've all lost our balance on a stairway before, so last week when you fell down the two stairs, were you rushing? Did you have your eyes and mind on task? Or were you thinking about something else?" And if the employee tells you—honestly—what really happened, then the person asking the questions needs to be able to help the injured employee work through which critical error reduction techniques and/or which habits he or she needs to work on. Finally, and most importantly, if the injured employee does tell you the truth about what really happened, it's critical that discipline is left out of the process. Otherwise, it won't take long before nobody is willing to discuss the states

☑ SAFE*START*® Incident Follow Up

RC _____
Date _____

❏ Personal Injury
❏ Vehicle Accident

Description of Incident:

_____
_____
_____

States → Which of the four states were involved in this incident?

❏ Rushing   ❏ Frustration   ❏ Fatigue   ❏ Complacency

Why

_____
_____
_____

Errors → Which critical error(s) increased the risk of this incident?

❏ Eyes not on Task   ❏ Mind not on Task
❏ Line-of-Fire   ❏ Balance/Traction/Grip

Why?

_____
_____
_____

← CERT What critical error reduction technique(s) [CERT] could have been used to prevent this incident?

❏ Self-trigger on the state (or amount of hazardous energy) so you don't make a critical error.
❏ Analyze close calls and small errors (to prevent agonizing over big ones).
❏ Look at others for the patterns that increase the risk of injury.
❏ Work on habits.

What can you do to improve this technique(s)?

_____
_____
_____

Figure 23
SafeStart Incident Follow-Up Form

☑SAFESTART®   Incident Reflection   **S** EDMONTON PUBLIC SCHOOLS **Skill**Centre

Student Name	Date	Time	Location/Rm

In reflecting on the incident that occurred with you, please identify identify which of the states applied and critical factors that affected this situation.

What States, Errors and Hazardous energy do you think may have been involved?

What Happened? (use SafeStart terms in your statement)

☐ Serious
 (hospital attention)
☐ Medical Aid
 (medical professional needed)
☐ First Aid
 (some in house treatment)
☐ Minimal
 (minor cut, band-aid)

What type of Hazardous energy may have contributed to the incident?
☐ Mechanical
☐ Thermal
☐ Chemical
☐ Electrical
☐ Other

Which of these four States may have contributed to the incident?
☐ Rushing
☐ Frustration
☐ Fatigue
☐ Complacency
can cause or contribute to these critical errors . . .
☐ Eyes not of task
☐ Mind not on task
☐ Line-of-fire
☐ Balance/Traction/Grip
. . . which increases the risk of injury

What Critical Error Reduction Technique(s) may have prevented the incident?
☐ Self-Trigger . . .

☐ Analyze Close Calls . . .

☐ Look at Others . . .

☐ Work on Habits . . .

Work On Habits
Critical Error Reduction Techniques
What habit(s) will you improve on to prevent a similar incident from occurring in the future?
☐ Test your footing or stance when you set up for work
☐ Look before your hand comes in contact with an energized equipment or material
☐ Move your eyes before you move your body
☐ Get your eyes back on task if you've been distracted
☐ Look for the line-of-fire potential before moving.
☐ Look for things that could cause you to lose your balance, traction or grip
☐ Glance up before standing up (to avoid banging your head)
☐ Keep hands out of pinch points
☐ Hold handrails on stairways
☐ Other

Student
Signature _____

Figure 24
SafeStart Incident Reflection Form

that lead to the critical errors. I mean, why tell anybody the truth if all it will do is get you into more trouble? Leave the discipline for deliberate rule violations—it's not really appropriate for errors such as slips and falls, or eyes not on task.

After we'd discussed enhancing accident/incident investigation awhile, I wanted to get back to some of the other elements in a safety management system. So I asked Gary, "What else would you suggest as a priority when integrating these concepts and techniques into a company's safety management system?"

CHAPTER 6—*Human Factors and Compliance Training*

Gary thought for a moment and then said, "Well, we've covered near miss reporting, JSAs/JHAs, risk assessment and accident investigation. So now we should start looking at compliance training. We all know we have to do certain types of compliance training. Some is done every year, like respiratory protection or hearing conservation; some is done every three years like forklift or process safety management; some of it is done on an as needed basis like lockout/tagout or hazard communication. Unfortunately, there's a tendency for organizations to just show the same video they showed last year. I've been in organizations where the same lockout video had been shown for 12 years in a row. Most compliance training is not very exciting. Employees don't really want to come; they're not paying much attention, and they're bored out of their minds. It's ironic that the training that's supposed to alert people to the hazards of the job and the proper work procedures to use to protect themselves is actually generating more and more complacency because the same old training materials are used year after year."

"I've seen that scenario played out as well," I said. "I don't know why they think they can show them the same video that they've shown them for years. You can imagine how much attention you'd be paying—even watching one of your favorite TV shows—if you were watching it for the sixth time in a row and knew exactly how the show ended. You're right; it would actually be fueling the complacency. Sometimes when nobody was paying attention, I've heard frustrated trainers say, 'Okay, we're going to test you on this afterward.' I guess employees might begrudgingly listen just to pass the test, and you could point to the test to prove that they were trained in order to limit your liability if anything ever happened. But what would happen to your defense if it could be proven that your training actually made employees more complacent; that instead of making things better, it made them worse? Training that claims to reduce liability but actually increases the risk of injury is obviously counterproductive."

"The best way to reduce your liability is to prevent the injury in the first place because then there is no liability; there is no question of who's at fault because nothing happened, and there is no damage to people, equipment or the environment. But there is a natural tendency for some managers to worry more about liability than they do about injury prevention: 'We have to do this refresher training. So, we'll get it done, and we'll get a check mark in the box,' as opposed to, 'How much mileage could we get out of this refresher training if we viewed it from an inside-out perspective?' In other words, they could say to the employees, 'We know you know how to lock and tag out this equipment.' 'We know you know how to fill out the confined space entry permit.' 'We know you know how to do all of the steps required for the confined space entry in terms of testing for atmospheric hazards, and so on and so forth.' So instead of telling employees what they already know—again—we should be asking questions like, 'What are the problems that complacency can cause with this task?' 'What are problems that rushing, frustration and fatigue can cause with confined space entry or lockout/tagout?' Getting the employees talking about the problems complacency can cause or that rushing, frustration or fatigue can cause with these tasks would be much better than showing them the same video again and again. So why not get these concepts and techniques worked into your compliance training so that it isn't a sunk cost that's actually making injury prevention worse instead of making it better?"

"I know that SafeStart has some PowerPoint presentations that accompany compliance-type DVDs," Gary interjected, "where you've integrated these concepts and techniques into the discussion. And don't you also have about 40 video clips of common workplace incidents where you highlight the states, errors and critical error reduction techniques (CERTs) that could have prevented the incident?"

"Yes, we do," I replied. "But I think the important thing to recognize is that you don't need a program to start integrating SafeStart into what you are already doing for compliance. If you conduct toolbox or tailboard sessions, all you have to do is just start discussing the possibility of rushing, frustration, fatigue and complacency when you talk about the work that's going on that day."

Gary just nodded. "What's next?" he asked.

CHAPTER 7—*Integrating Emergency Preparedness and Workplace Procedures*

"Maybe without going through all the rest of the elements in detail," I told Gary, "you could at least give us an idea of how you integrated these inside-out concepts into some of the other elements in the safety management systems at the companies where you've worked. One of the elements you've mentioned before (I've heard you talk about it many times) is emergency preparedness. Whenever you've moderated a session that I've been at, one of the first things you tell the people, among all the other housekeeping issues, is where the emergency exits are. Often you ask them if they have fire drills at work. They all put their hands up. Then you look at them and say, 'Do you get everybody out?' After most all the hands drop and everybody starts looking sheepish, you say, 'I know you don't, so don't tell me you do. I know you don't get them all out because we used to have fire drills all the time, and I know we didn't get everybody out. Sometimes the worst culprits are the people who run the place.' So maybe you could talk a bit about emergency preparedness."

Gary smiled. "I can talk a fair bit about emergency preparedness and fire drills because it's one of my favorite topics. What you said happened at the sessions is exactly what happens," he replied. "We'd have a few people who insist that they get everybody out, so I'll ask, 'How hard is that to do?' They'd tell me they actually have to go around and personally try to get them out of there. Quite often it's the general manager or the plant manager that's the last one to leave because they probably know that it's a drill. Everyone in the plant knows you're going to have a fire drill every year on every shift; they know the drill is coming. So, when the fire drill comes, not everybody gets up and leaves because they're not too excited about this. After all, the fire drill will cause them to have to stop work; who knows how long it will take, and they've got all kinds of work to get done. So they don't want to leave. The tendency is that they also don't believe there really is a fire. Eventually you might have to force them all out. Unfortunately, human nature being what it is, the more drills you have, the more likely it is that they are going to think it is just a drill and become complacent. Then when you do have a *real* fire, they might not leave if they don't smell the smoke or see the flames. So you can actually train people to become complacent. Now I'm not saying you should stop obeying the regulation and cancel all fire drills, but you've got to take these human states and errors into consideration during training."

"Complacency happens to us all. I was doing some work in Kitchener, Ontario, and I was staying on the tenth floor of a hotel. Lo and behold, at one o'clock in the morning, the fire alarm went off, and of course, my first thought was that some kid had set off the fire alarm. But I got up anyway and walked down the ten stories because the elevators were shut down when the fire alarm went off. There were quite a few people going downstairs. As we were all milling around outside, the fire department arrived, but couldn't find any fire. So they allowed us back in the building. Unfortunately, they didn't turn the elevators back on, and I had to walk up all ten stories."

"About half an hour later the alarm went off again. I wasn't too keen on getting up and going down since I've got to work the next morning. As it was I had to get up in a couple hours anyway, and I wasn't too happy about it. But I'm a safety professional; I feel obligated by my profession, so I got up and walked downstairs— the same scenario all over again. However, this time there were a whole lot less people outside the hotel than there were the first time. The first time there were a couple of hundred, the second time there couldn't have been more than fifty. The fire department came and, once again, they still couldn't figure it out, so they let us go back in. I traipsed up ten stories and tried to get some sleep."

"Half an hour later, the alarm went off for the third time. Now, I am really thinking seriously about not going down, but I do anyway. This time there weren't more than a dozen people outside the hotel. The fire department was disgusted with the whole situation, and they were walking around the building trying to figure it out. I knew that the hotel had an automated alarm system. I also knew that this type of system did not have many false alarm issues. So since I was already up, I decided to walk around the building to see if I could figure out what was making the system think there was a fire. In the back of the building I found a small, smoldering fire in a dumpster. Evidently, there was just enough smoke getting into an air intake of the building that it was setting off the automated system. So I told the fire department, and they took care of it immediately."

"Since the situation wasn't much of a risk, what's the point of the story? The point is that there weren't even a dozen people who actually walked down the stairs to exit the hotel when the third alarm sounded. Now think about the workplace. If you keep crying wolf or yelling that the sky is falling, you are actually training people to be complacent. Then when something really, really serious happens, they won't respond in the way that you think you have them trained to respond. It *will* happen over time if you don't start integrating human factors into the training so that people are able to react appropriately. You need to ensure that they understand that it isn't just, 'Oh, we have to go through the motions and get our signature on a piece of paper because that's what the government says we have to do.'"

I looked at Gary. "That's a great example of how people respond when they have to repeat something like a fire drill over and over again," I said. "Nobody is really thinking about the complacency factor as these drills go on over time. What will the results be in five years? Will people respect the drill, or will people have become conditioned to disregard the fire alarm? That's an issue that's important from the company's perspective."

"But I've also seen complacency with emergency preparedness from an individual's perspective many times. Out in the oil patch, for instance, everybody wears an H_2S detector, and there's good reason for having one. Hydrogen sulfide is one of the most dangerous, toxic substances out there. In spite of how dangerous H_2S is, it was very common to find people who had taken the batteries out of their detector because they needed to use them somewhere else. I've also seen this happen with a missing attendant or safety watch during a confined space entry. 'Where is the safety watch?' 'He went to get coffee for all of us.' 'Why?' 'Because we wanted some coffee, and he never does anything anyhow.'"

"So we want people to be looking at those four states since they can be found in every situation, even emergency preparedness. Complacency can interfere and cause problems getting everybody out of the building during a fire drill. But it can also affect personal decision making, allowing you to render an H_2S detector inoperable or putting your co-workers in harm's way by going for coffee."

But there's more, so I asked Gary, "What about workplace procedures that involve personal protective equipment, machine guarding, or even something like lockout/tagout?"

Gary had a quick answer. "When you think about PPE from

an inside-out perspective rather than an outside-in perspective, everything changes. Outside-in thinking is, 'Okay, we've got a hazard; we need PPE to take care of it, so we're going to buy the cheapest, most efficient one that is going to do the job.' But from an inside-out perspective, we need to think about this differently. We don't want to necessarily get the cheapest one that we think is okay because it may not be the most comfortable. The comfort issue is just as big an issue as effectiveness. It's frustrating when you have PPE that isn't comfortable and doesn't work properly. Yes, you might have to spend a buck or two more, but if the cheaper one is not comfortable or if it's difficult to use, your employees get frustrated with it and may not use it. Think about a pair of safety glasses that are uncomfortable and continually steam up. Not only are the glasses not comfortable, they don't allow you to see what you need to see (I hear this from maintenance people all the time), and the next thing you know, the glasses are off because the person can't see what they're doing. The fact that the glasses don't work right gets the employee frustrated, and then the employee is willing to increase their risk by not wearing the glasses."

I told Gary that I knew what he meant. Then I gave him an example of a person who works in a machine shop and uses an air wand to clean a lathe or a milling machine. The safety procedures require the use of goggles in addition to the safety glasses to prevent little pieces of metal from blowing up under the safety glasses and into the eyes. This can increase frustration because goggles are more uncomfortable. They make you hot and sweaty, and if you don't have a fan or an air conditioner in the area, then the goggles and/or the safety glasses will fog up. The frustration builds when you've got to see precision work where the tolerance is a couple of thousandths of an inch (0.005 centimeters) , but you can't see because your eye protection is all steamed up. The frustration doesn't get any less when you tell your manager or supervisor about the problem and they tell you, "I don't care! You know what the rules are; you have to wear the goggles."

"They've designed frustration right into the system," said Gary. "Unfortunately, not every manager sees frustration as a serious problem."

> I knew it would take a while to get some safety professionals to treat rushing, frustration, fatigue and complacency with the same respect as they would a physical hazard. I also knew that it would take even more time before most managers and supervisors would treat these four states with the same respect. They would also need to recognize how some of these things interface. For instance, if you look at the concept of lockout/tagout, we all know we have to do it. We all know it's a legal requirement. We all know we have to train people on it; but how much effort are we expending to making it easier, less cumbersome and less time-consuming to lock something out? I think everybody would agree that the more difficult and time-consuming it is to lock and tag something out, the less compliance we're going to see. So I asked Gary if he'd seen anything like this.

"I'll give you a good example," he said. "At one of our facilities we had a washer system. All the material and all the weldments that were going to go into the end product had to be washed through this system before they were dried, and then undercoated and painted. The washer system was important; it was the heart of the production system. There were eight pumps that pumped the water through the washer system. If one of the pumps failed and needed replacing, it took 18 locks and a lot of time to lockout the system. Replacing one pump meant shutting down the entire washer system *and* the entire production system. With the entire production line shut down, everybody would get upset. We constantly had lockout violations caused by people trying to get a pump replaced quickly. Both supervisors and employees just looked the other way, rule or no rule."

"When I went out to look at it and to talk with the operator and maintenance guys to see what could be done, the first thing I asked was, 'How often do these pumps go out?' They told me that one goes out about every week (the pumps were under a lot of stress). 'So if one pump goes out and we only have seven pumps operating, what impact does that have on the quality?' 'Actually, none. We can probably get by with seven pumps, but we wouldn't want to go down to six.' 'Okay,' I said, 'so what if instead of having 18 locks to lock this out and having to shut the whole system down, what if we set it up so that we could guard it properly and only had to lock out one pump? Then we could take a few bolts out, slide the old pump out, set it aside on a cart, bring a repaired pump motor in, set it down, put the bolts on, and slide it back in place so we wouldn't have to shut down the line at all. The lockout would only require one lock.'"

"So we were able to redesign the equipment and revise our lockout/tagout procedures so that if we only need to replace one pump, the washer system and production system could still be running. We put some protections in place so that the employees doing the pump repair couldn't get into anything that would hurt them, so that to replace one pump, only one lock was required. The procedure that required shutting down the entire production line to replace one pump caused lots of rushing and frustration. The old procedures we used provided enough irritation to motivate people to take shortcuts. The change to one lock took a lot of the frustration out of the equation."

"And you took some of the rushing component out, as well, because the new lockout procedure just took less time," I said.

"That it did," Gary said. He paused briefly, and then said, "We've spent quite a bit of time talking about how to integrate these concepts into certain elements of a traditional safety management system, but one element we haven't talked about yet is job task observation. Since you've been involved in behavior-based safety for years, tell us how to integrate these concepts there."

CHAPTER 8—*Integrating Traditional Behavior-Based Safety*

I looked at Gary as if to say, "You're doing it again. I'm supposed to be interviewing you, remember?"

He just looked right back and said, "Hey, nobody's made more observations with more workers in more industries than you have. So, yeah...you should be the one to explain this."

"Okay, okay," I paused to collect my thoughts. "Well to start with, although not every company has an element for 'job task observations' in their safety management system, most of the well-performing companies do."

"Some companies prefer to use 'complete' observations where someone (usually a supervisor) observes an entire job being done according to a set procedure. If they're using this method, then hopefully the observer is using a revised four-column JSA, or at least an updated standard operating procedure that includes human factors (states) and critical errors."

"Other companies prefer 'spot' observations where someone makes a brief (usually only a few minutes) observation of a co-worker, instead of looking at the entire job. They use a checklist of critical behaviors to see if the person is following those during the observation period. These are the kind of observations you usually find in behavior-based safety processes. In other words, if the company has an observation and feedback process, they are doing spot observations."

"So if the company has an observation and feedback process, whether it is supervisor to employee or peer to peer, you can enhance its value by adding the states and errors that you can objectively see to your list of critical behaviors. For example, you can objectively see rushing. You can also objectively see eyes not on task and body position out of the line-of-fire. You can also add things associated with balance, traction or grip. You can objectively see if somebody loses their balance or not, and you can certainly see things that could cause them to lose their balance, traction or grip. So there are definitely things you can add to your observation process (see Figure 25). However, just because you can't objectively see some of the other states doesn't mean that you can't talk about them."

"Rushing is something you can objectively see, but the other three states are much more subjective. You can't always see frustration. Sometimes it's very visible, but other times it's hidden below the surface. Sometimes you get to see signs of fatigue, but it's difficult to gauge how tired they really are. Just because you see somebody yawning doesn't mean they're too tired to drive a fork truck safely. That would be something you'd need to talk to them about during the feedback part of the observation. Finally, we all know you can't see complacency objectively, but you can see evidence of complacency. For instance, if somebody isn't willing to use a safety device such as a fall arrest harness, it's probably because they don't think they're going to slip and fall. But even if a person is wearing a safety device that does *not* mean that they are not complacent. So, unless you see something really obvious, it's hard to tell how complacent somebody really is just by observing them."

Figure 25
SafeTrack Observation Card

"Even though some of the states aren't objectively observable and you wouldn't want to put them on the checklist of critical behaviors, that doesn't mean the observer can't bring them up as part of the discussion. For instance, the observer could ask, 'Which of these four states is the worst for you: rushing, frustration, fatigue or complacency?' If the person says complacency, you could ask several different questions: 'Okay, on a scale of 1 to 10, how complacent do you think you are today?' 'How complacent has this job made you over time?' 'How often do you think about the risks?' 'How often do you think about the major risks?' 'How often do you think about some of the minor hazards?' You can also certainly talk to people about the pacing of the job and the level of frustration they might have. So while we can't objectively observe for all four states and all four errors, we can still get at them. We can certainly make sure we're covering them in the feedback or discussion part of the observation."

"So, if one of the elements in your safety management system is a behavior-based safety process or an observation-feedback process, it's definitely worth integrating these concepts. This will do a number of things for you, but one of the most important is that it's going to break into the complacency cycle (see Figure

26). Over time, the level of a person's awareness—eyes and mind on task—tends to go down. For example, when you first start something, you're a lot more aware; your eyes and mind are on task. But after a while, you can walk without necessarily paying too much attention to it. You can drive without paying too much attention to it."

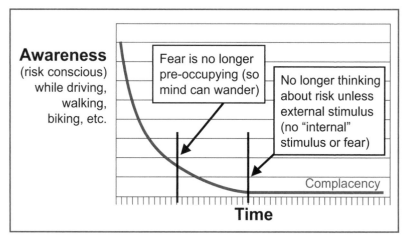

Figure 26
Complacency: Awareness Over Time

"There are two distinct points in time, labeled on the chart, that differ for each individual. The first line on the left represents the time when you stop paying moment-by-moment attention, when you've done an activity or a job long enough so that the fear is no longer preoccupying, which means your mind can wander. It also means you're now very susceptible to rushing, frustration and fatigue. In other words, you don't have the risk associated with the job to help focus your mind on the task, so any of those psychological distractors will make you much more likely not to be thinking about what you're doing. For instance, you're no longer preoccupied with the risk, but now you're in a bit of a rush. You'll most likely be thinking about why you're rushing or what will happen to you if you're late or what will happen if the job doesn't get done on time. So the first point on the chart is where you're no longer preoccupied with the risk so your mind can wander."

"The second line to the right represents a longer period of time for doing the same job or task. When people get to this point, they are no longer thinking about the risk *unless* there is an external stimulus. Driving your car is probably the best example for this. Nobody that I know gets in their car to go down to the corner store and says, 'Oh my gosh, if I just focus, if I just concentrate, if I really pay attention, I can make it the three blocks. I can park the car. I can get out. I can buy the milk and the bread and the sugar, get back into the car and, hopefully, if I really pay attention, I'll make it back home again without being seriously injured or killed.' Nobody does that. We all know, though, that most car crashes occur near home. In fact, almost 70 percent occur within 10 miles (16 kilometers) of home.[3] So at this line on the chart, you're not going to be thinking about the risk before you get in the car. But if on your way to the corner store somebody blows through a red light and nearly hits your car, well now, you're going to be thinking about the risk. That's only because there was an external stimulus. Likewise, if you saw a bad car wreck on the side of the road or a car wreck in front of you, you'd be thinking about the risk. Quite frankly, even if you see any other kind of a state-to-error pattern and you recognize it, you'll start thinking about the risk. However at the second line, overcoming complacency by getting your mind to focus on the risk requires that external stimulus."

"An observation and feedback process provides that external stimulus without increasing the risk. What you're doing is breaking into that complacency cycle, and...I don't want to use the word 'forcing,' because it sounds like you're trying to control people...but you're forcing the issue, forcing the attention back to safety and risk. You're starting to ask people: 'What parts of this job could you become complacent about?' 'What are the

---

[3]CarInsurance.com, "Are Accidents More Likely to Occur Close to Home?" < http://www .carinsurance.com/Articles/content31.aspx>, accessed April 9, 2012.

worst hazards?' 'What are the less obvious hazards?' 'How do you protect yourself from them?' 'What would you tell somebody you're training to do this job about what the worst hazards are and what the least hazards are?' 'If you could only tell them one thing, what is the most important thing you would tell them?' 'If I asked you what is the worst close call you've ever seen, either with yourself or somebody else doing this job, what would you tell me?' All of those questions are designed to get people refocused on the risk of the job, where after five or ten years they probably don't think about it at all, any more than we think about the risk when we get in the car."

"We can't afford to have near misses motivating people to focus their attention back on the risk. We can't rely on serious injuries to focus people's attention back on the risk. We certainly know it happens when there's a fatality in a facility. Everybody in that facility is less complacent for a while after that. But that is an extreme event that you can't rely on, nor do you want to have happen. But a simple observation can provide that external stimulus and help to fight complacency. And that, by the way, is why I think behavior-based safety processes really do reduce injuries versus the ABC theory. So you need to get these four states and these four errors integrated into your observation and feedback process, and you also need to eventually get them worked into all of the other elements in your safety management system."

"Okay before we get into some other issues," I continued, "I just wanted to review your suggestions for integrating the concepts and techniques into the safety management system. So the first thing you'd suggest is near miss reporting because that's going to prevent a lot of injuries quickly."

"Right," Gary said.

"The second thing would be working on your job safety analysis so you get the concepts and techniques to be a part of the *fabric* of the system and create better risk assessments."

"Yes, now the concepts and techniques are in black and white, and harder to be forgotten," Gary replied. "It's harder for SafeStart to become a flavor of the month."

"The employees have to be involved in this process too. When they're involved in identifying the potential problems that rushing, frustration, fatigue and complacency can cause, it will improve the whole process of risk assessment, both from a company perspective, and an individual one as well. Now they're focused on risk assessment from an inside-out perspective. They recognize that since the majority of risk is in the 'self' area, they need to assess that risk first and foremost, instead of last, or not at all."

"Right," Gary said.

"Once you do that, if everybody's 'cool' with the first two steps," I said, "and a majority of the employees have been taking SafeStart home to their families and using *Hurt at Home* with their kids, the third step you'd recommend is enhancing the accident/incident investigation process."

"If parents are using SafeStart at home with their kids," Gary said, 'and talking to them about how they got hurt and why, that will minimize any backlash at work when the accident/incident investigation is enhanced. It's important for employees to realize that SafeStart isn't about punishment or discipline. It's about helping to improve our personal safety, not a blame game."

"The fourth priority step would be to work in the four states, the four errors and the four critical error reduction techniques into your compliance training," I said, "so that the training doesn't become a waste of money by promoting complacency."

"There's nothing more counterproductive," Gary replied, "than looking at the same video year after year and hear employees say, 'Oh no, not this thing again!' It just makes them 'turn off' and become more complacent."

"By integrating these concepts, you know, doing it from an inside-out perspective," I said, "you actually start to get a return on that investment and improved compliance, as well as improved injury prevention. By listening to a story about someone who entered a confined space or failed to use lockout/tagout because they were a little complacent or in a rush, the safety awareness of each employee is heightened, and compliance training becomes motivating rather than demotivating."

"Right again," Gary replied.

"There are many more elements in a safety management system than the ones we've mentioned here. These are just the first four you might want to look at in the beginning."

"Of course, then you've got to start looking at the other elements of the safety management system: emergency preparedness and workplace procedure right through to your observation and feedback process. But if you start with these four first, you'll find that the rest will fall into place much easier."

When we finished reviewing our priorities for integrating the SafeStart concepts and techniques into the safety management system, I decided it was time to take a broader viewpoint in terms of reengineering and rethinking the workplace from the inside out.

I asked Gary, "If we were just looking at the four states, what can you do in terms of redesign and rethinking what you're doing from the perspective of rushing, frustration, fatigue and complacency? I mean, can you design things to make people less complacent, or design things to take complacency into account? You gave us an example earlier about a situation where you were able to redesign a system so that when you needed to replace one pump, only one lock was required instead of the original 18 locks. What other examples can you give to illustrate rethinking the workplace from the inside out?"

CHAPTER 9—*Using Inside-Out Thinking to Reduce Rushing, Frustration and Fatigue*

"When you finally get your mind around the idea that you are not only looking at safety," Gary began, "but also productivity, quality, and managing process costs, then it's essential to get the leadership group to change their perspective from outside in to inside out. When they do, they'll start asking questions like 'What are we doing to create situations where people have to rush?' 'What are we doing where we're creating situations in which people are frustrated?' 'What are we doing to create extreme fatigue in people?' Let me just give you one practical example, and then we can go from there, because there are all kinds of opportunities out there."

"A large company had an operation to weld frames for air conditioners. These frames were 12-feet (3.7 meters) wide and 20-plus feet (6-plus meters) long. All of the metal framing parts had to be laid out, clamped and welded in place in a booth using two MIG welding machines, one at either end of the long booth. Because of the sequence in which the frame had to be welded, the welder would have to start welding with the gun from the MIG welder at one end of the booth, stop welding, put that welding gun away, walk down to the other end of the booth, get the welding gun from that MIG welder, stretch the hose out and start a new weld at the other end of the frame. When the sequence called for him to weld on the other end of the frame, he'd have to reverse the process: put the welding gun away, walk down to the other end of the booth, get the welding gun from that MIG welder, stretch the hose out and start welding in a new position. Some of the welds were so long that he couldn't make a continuous bead using one MIG welder. He'd have to stop, put the gun away, walk down to the other MIG machine, and finish the weld using that gun."

"Although the welder was a hard-working man who loved what he did, all of this constant back and forth and back and forth put a lot of excess mileage on his legs and really tired him out. It also affected the quality of the welds. No matter how good a welder someone is—and he was a good one—you can tell where one weld stopped and the other one started. It leaves a seam that you don't want, and he'd have to cut that out and start the weld all over again. All the trouble the welder experienced repairing the seam and making sure it was done really, really well, caused him to become frustrated. To complicate the situation even more, the welder had a quota for the number of frames he had to weld each day. His job was on the line if he didn't produce. So every time he had to switch welding machines or had to repair a weld, every second he was delayed increased his frustration level and caused him to rush to make up for lost time. All this rushing also affected quality. It was a vicious circle."

"The welder wasn't just hard working; he was also smart. He came up with a way to eliminate a lot of the rushing, frustration and fatigue. 'Why do we have a welder on one end and a welder on the other end?' he asked management. 'That setup won't allow me to make a continuous bead on some welds. If we put one welder along the side of the booth on a swing boom I could run an entire weld right straight through, and I wouldn't have do all this running back and forth.' He made that suggestion numerous times. Finally, somebody saw the value of his suggestion and implemented it. It made a difference—a big difference. It improved his ability to get the job done in a timely manner without rushing. It solved the quality issue that had caused a lot of frustration when he had to dig out the weld and start all over again. Unfortunately, it also had an unforeseen negative consequence. After a time study of the new setup was complete, management told him, 'Well, you can do this job faster now, so we're going to expect you to finish more frames each day.'"

"I mentioned this story for two reasons. First, is it possible to reengineer a job to make it easier? You bet it is. When you use inside-out thinking, you can take a lot of rushing, frustration and fatigue out of the system. But second, you can add rushing, frustration and fatigue back into the system just as easily. That can happen when you neglect to account for the effects of crucial parts of the system, in this case, the time study. We can't engineer every hazard out of a task, but we can use inside-out thinking to improve productivity, quality and safety, and engineer out most of the things causing rushing, frustration and extreme fatigue."

As I listened to this story, I could certainly see it from the welder's point of view: "I made an improvement. I made things easier. I improved manufacturing efficiency, and as a result, instead of a raise, I got a higher quota." In the end, how much did this alleviate the welder's frustration level? It probably made things worse, even though on the surface it would look like a lot of things had improved. This reminded me of a situation that Gary and I knew about where we thought they would have known better than do what they did. It involved a manufacturing facility where they'd given people training on how to better cope with rushing, frustration and fatigue. They'd taught them the self-triggering technique, and, if you listened to their opinion, they'd taught them well. Then, after the training was finished, they decide to cut back 10 percent of the workforce and speed up the line.

Gary is sort of our main troubleshooter when things don't go well. Thankfully, we don't have to send him out every week for things that aren't going well, but once or twice a year, he has to go in and fix up a failed implementation. I asked him to talk about that manufacturer. Although it's an extreme example, it illustrates the danger of rethinking the workplace from an inside-out perspective and then doing something, without thinking it through, that is the exact opposite of what you should do.

"This one was really a bit frustrating," Gary said. "It was a major company, a Fortune 500 company, respected throughout the world for their excellent products. One facility in particular was giving corporate a little bit of trouble. They were not performing very well in the safety area even though they had an observation process (supervisor to employee) in place and had some very competent, actually marvelous, safety engineers. They decided that the thing they needed to do was train everybody with SafeStart. For this implementation we trained their trainers on how to use the SafeStart materials, and then their own trainers trained their employees. Normally we see injury reductions almost immediately. However, in this particular facility we actually saw increases in injuries—and they were significant. No safety training in the world should ever create injuries. So, I got called in to talk to the management to see what was going on. They basically told me that they did the training right, and they wanted to know what *we* were going to do about it. After a few more preliminaries, I said, 'Let me go out on the floor and talk to some people, tour around and see what has happened.'"

"I walked down one of the assembly lines, but didn't get much right off the bat. Although the employees had never seen me before, they knew I was with SafeStart; someone must have told them. As I continued down the line I talked with one of the guys. He told me, 'Well, you taught us not to rush, and now we have to rush.' I said, 'Tell me about that.' He said, 'We used to have a large number of people in this line' (I think it was sixty) 'and they decided to re-time the jobs after the SafeStart training. Now we have about 50 to 55 people on the line. And not only that, it's an automated line, and they've sped it up 10 percent.' Now, I have to admit that when I looked at this line before, I was amazed at how inefficiently the line ran and how little the people seemed to have to do. That was prior to the SafeStart training. From my experience, I would have expected more. I don't think the employees were being overworked, and I'm not so sure the company was getting their money's worth. Evidently, the company had figured this out too. So right after the SafeStart training, they sped up that line 10 percent and cut the manpower 10 percent. They made adjustments on the other assembly lines in the facility too."

"The production changes did not happen because of SafeStart. They just happened to come immediately after the SafeStart implementation. But what they did and when they did it caused a whole bunch of frustration—bordering on anger. Employees felt like, 'You tell us not to rush in the training. But the very next thing you do, within two weeks, is speed up the line and cut our manpower. Now we *have* to rush.' It became a self-fulfilling prophecy, and the injuries went up. I don't know whether they were contrived injuries or whether they were real. But management did exactly the wrong thing, and it really created a situation of unrest that led to a lot of rushing and a lot of frustration. I would have been surprised if injuries hadn't increased."

I thought Gary's story pointed out the fact that there is such a thing as "legitimate" rushing: They sped up the assembly line so an assembly line worker would have to move quicker. That's fairly straightforward. But management failed to implement the change incrementally so that the assembly workers could get used to the quicker pace. In other words, it doesn't matter whether the rushing is "legitimate" or not, it still leads to errors and injuries unless you do something to mitigate the effects. I remember working at an oriented strand board (OSB) mill in the middle of nowhere. It's a good place for an OSB mill because the trees typically are all out in the middle of nowhere, and it's expensive to transport them. So, you put the mill where the trees are. The whole plant floor was painted with a non-slip Epoxy coating so you weren't walking on something as slippery as a skating rink. As you'd imagine, this paint wore off fastest in the high-traffic areas. One of the first places where the paint wore off was a 90-degree turn into the lunchroom from the walkway at the side of the large OSB press. Recently the lunchroom was equipped with one of those coffee machines that brew individual cups of coffee. It makes a great cup of coffee, but it takes a little bit longer than just pouring a cup out of a pot.

The maintenance department at this particular place just happened to be at the back end of the plant, beside the shipping area. So if you worked in maintenance, in order to get to the lunchroom, you had probably the longest walk of anybody in the plant. You had to walk all the way across the plant and then almost all of the way down the length of the plant to get to the lunchroom. To complicate the situation, the plant had a rule that you could not smoke on the property. (The plant manager didn't smoke, and he was trying to promote wellness.) That meant if you wanted to smoke during your coffee break, you had to get into your car or pickup truck, drive across the tracks to the main road, and then when you were finally off the property, you could have a cigarette.

Since a lot of people want coffee and a cigarette at the same time, you can probably see the problem here. The maintenance guys had the longest walk, and they had to get in that line real quickly if they wanted coffee, because just the line for the coffee might take five to ten minutes. Then they had to get into their car or truck, drive across the railroad tracks, have a smoke and make it all the way back to where they were supposed to be working, in the fifteen minutes that was available. I'm not saying people were moving at a dead run—I don't think you ever see maintenance people on a dead run unless there really is a fire—but they were rushing. I was concerned that this 90-degree turn didn't have any non-slip paint, and it was going to cause somebody to slip and fall. So, I explain all of this to the plant manager, telling him that we have an accident waiting to happen because people are going to be inclined to rush when they want to get outside to have their coffee and cigarette, and the 90-degree turn is one of the most slippery places in the plant. His response back to me was, "Well, they shouldn't smoke."

I told Gary that eventually they did repaint this 90-degree turn with non-slip paint. Then I asked him, "Do you think they did this before or after somebody slipped and fell, hurt their wrist and had a recordable injury?"

"My guess is that it happened after the injury," Gary said. "But you're right, it doesn't matter whether they should or shouldn't be rushing. What really matters is that if you reduce the rushing, you reduce injuries. One of the companies where I was a safety manager had a huge facility on 1,800 acres with multiple buildings spread out over it. However, there was only one cafeteria, and it was at one end of the plant. We were having a significant number of slips, trips and falls as people were walking from one building to the other. I first thought they were mainly caused by snow or rain, but when I started looking closer I found that a lot of them happened in good weather when someone just stepped off the sidewalk and twisted their ankle. I also found out that most of the slip, trip and fall injuries tended to be clustered between 11:00 a.m. and 1:00 p.m. How could I have missed that? It didn't take a rocket scientist to realize that this must all involve people rushing to the cafeteria. You see there was only one on campus. You could get coffee in other areas, but if you really wanted a meal at lunchtime, it would be a long walk, and you'd have to walk pretty rapidly."

"But I wanted to confirm my hunch with a little bit of observation. I looked inside the cafeteria first. The cafeteria was designed wonderfully so that you didn't have to stand in one line and experience a lot of delay. There was a deli line, a soup line, a hot meal line and a drinks line. So you could move through the cafeteria quickly. Then I just sat out front and watched what was going on. I realized that while employees weren't running from one building to another, there was a lot of traffic headed for the cafeteria, and they were all moving at a pretty good clip. Some people had half an hour for lunch; other people only had twenty minutes."

"So I approached management and said, 'I think we ought to think about adding another cafeteria at the other end of the campus where employees can also get a hot meal.' At first there was a great deal of resistance to the idea, but I reminded them that we already knew we had to make some changes, for example, adding more women's restrooms and locker rooms to accommodate the increased number of women working in the facility. I also reminded them that that end of the facility didn't have any training rooms, and we were paying people for walking a long distance to attend training and then walking back. I wondered if we could save some money if we included a small conference room, a couple of training rooms, along with a small cafeteria. Amazingly, when we actually added those facilities, we took some risk out of the system, and our slips, trips and falls dropped to hardly anything."

Hearing him, I realized this wasn't a cheap solution. However, what they were doing was not just fixing the problem piecemeal—they were eliminating it. When you look at the savings in terms of those slip and fall injuries over a ten-year period, my guess is that they probably saved a fair bit of money. So I asked whether this was eventually a profitable decision.

"Yes it was, and obviously we got a lot of good PR out of the deal," Gary answered. "But what I think is most important is that if we hadn't been thinking about the state-to-error pattern, if we'd only thought about things from a traditional production, quality, cost standpoint, we would have missed this opportunity. We had to look at the human factors or we never would have figured out that adding a satellite cafeteria would eliminate the major cause of the slips, trips and falls injuries we were experiencing. Adding the training rooms made the whole project cost effective since we reduced the time people spent getting to training, so that they were away from their work a lot less. I would not have wanted to face my boss if the project hadn't worked out, but it panned out very well."

When you look at designing the workplace to reduce fatigue—the immediate fatigue of doing a task, over and over again, whatever it is, whether it involves your back, your wrist or your shoulders—ergonomists have made great improvements in the last twenty years redesigning machine centers and workstations so workers don't have to reach over so far and, as a result, don't have to strain their shoulders and backs as much. So there has been a lot of improvement in terms of specific muscle fatigue.

There's also a lot you can do to negate the overall level of fatigue at your workplace. If you run shifts, you probably know that 20 percent of the workforce copes with shift work pretty well, 20 percent copes so badly they quit within the first couple of years, and 60 percent of the workforce copes day to day. (They're thinking, "If I can just get through today....") Companies could reduce a lot of the fatigue by focusing on the way they run and rotate shifts. They could certainly train their workers on how to cope with the shiftwork better. This would also help with the overall amount of fatigue.

I remember when my wife worked as a nurse at a large, metropolitan hospital. The hospital was running 12-hour shifts, and the typical rotation for the nursing staff was to work two days and two nights, or two nights and two days. What they were trying to do was to split up the work so that everybody had to work the same number of dayshifts and the same number of nightshifts. But if you try to work two dayshifts in a row, and then take a day or a couple of days off, and then work two nights, or worse, work two nights and then get up and work two days, your body just can't recover fast enough. The nurses who worked in the Level II nursery dealing with premature babies followed the same rotation. Just imagine trying to get an intravenous needle into a premature baby when you're really tired; it's like hitting a needle with a needle. Now, try to imagine doing it drunk, because that's about what happens to your fine motor skills when you're really tired.[4] The nurses would be sticking these little babies 10 or 12 times trying to get the needle into a vein. It wasn't a skill problem. It was a fatigue problem—they were just too tired. Teaching these nurses how to cope with shiftwork better and how to pick their shifts better would have gone a long way.

Of course, some companies have recognized the problem fatigue causes. They have taken a closer look at the non-standard hours they require their employees to work, and have done what they could to minimize the negative mental and physical effects of shift work, including shift lag. Some employers have set aside special rooms where employees can take a 20-minute nap. They've trained their employees about circadian rhythms and strategies for dealing with the impact of shift work on their personal lives. They've even included the family in the

---

[4]Drew Dawson and Kathryn Reid, "Fatigue, Alcohol and Performance Impairment," *Nature*, Vol. 388, July 1997, <http://www.fatiguescience.com/assets/pdf/Alcohol-Fatigue.pdf>, accessed April 9, 2012.

training, so that everybody understands how important it is that mom or dad gets the right amount of sleep during the daytime, and that two hours plus two hours plus two hours plus two hours doesn't equal eight hours of sleep.

Unfortunately, at other companies, it's still, 'If you fall asleep on the nightshift shift, you're fired.' Instead of encouraging people to get naps and deal with the fatigue head on, they're doing just the opposite. Reducing fatigue in the workplace is a two-step process. The first step is looking at the job task from the point of view of how it could be tiring out somebody's muscles, or how it might be tiring out or overworking somebody's shoulders or back. But the second step is equally important: looking at what the job task does to the overall level of fatigue, especially if you run shifts.

I asked Gary for an example from a traditional manufacturing environment where they had done some things to design out fatigue, but not necessarily from an ergonomist's point of view.

"Shiftwork is a big issue," Gary said, "and it's going to get even bigger. The reason we have shiftwork is to maximize the utilization of capital (the equipment and the machinery) over a 24-hour period of time rather than only 8 or 10 hours a day. For example, you might have a 2-shift operation over a 24-hour period of time. The same drive to fully utilize capital also motivates companies to fully utilize manpower. Some processes, like generating electricity, are very difficult to shut down. The same is true for steel mills and pulp and paper plants. In these environments, you'll see 12-hour shifts, rotating 12-hour shifts, or even what are called 'weekend shifts' and a number of other unique scheduling methods to maximize equipment and manpower. In some instances, we've tried to reduce fatigue by increasing the light at night so that the body thinks that it's daytime instead of nighttime. Some workers just can't cope with shift work, and they 'self-select' or quit in order to find other work in the same facility that does not require it or, if that's not possible, go somewhere else to work."

"Unfortunately, some work rules kind of force fatigue onto the system. That's true of how overtime is administered in most union facilities. When there's a need for overtime, quite often there's a requirement to offer overtime to the employee that is currently doing the job. So let's say we need an employee who works 7:00 a.m. to 3:30 p.m. to work over. Instead of being able to have somebody else do the work or divide it up into smaller four-hour increments, we have to offer the overtime to the employee that is just finishing up his or her shift. We also have to offer the entire eight hours to them. So now you have a person working 16 hours and then driving home. Then they have to get up and drive back to work the next day without having had enough sleep. What this does is to mess up two different work shifts: the shift they worked as overtime and the next day's shift. Chances are they probably were not in very good shape for either. But it's rare that anybody turns it down. After all, when you're talking about time and a half or double time, the money's a big deal. And if they did turn it down, they are potentially losing an opportunity to work overtime the next time it's available. So it's now or never from the employee's point of view. And to make matters worse, if they worked overtime one day and the next day there was still a need for overtime, they might take the whole eight hours if they needed the money or maybe *now* they'll tell you they'll only work four hours because they're exhausted. The point is that the system drives people to accept overtime that they may not really want."

"We know they're not as productive because production studies indicate that you can only work about ten hours before your efficiency and your quality diminishes; you're not as productive. We're ending up paying more than full wages for somebody who isn't going to be productive. We really ought to be thinking about how we handle the overtime situation up front in our contract negotiations and find a different way to allow people to work a reasonable amount of overtime when it's available. We should not be setting employees up so they have two or three days in a row where they're inefficient, they're really, really tired and they're increasingly prone to mistakes: mistakes that cause injury and mistakes that affect quality."

It's interesting, I thought, how trying to be "fair" can actually exhaust people. But even worse, now the exhausted employee probably has to drive home. That's a huge risk. That got me thinking about the last state, complacency. Can you design your workplace for less complacency? Or, if that's not possible, can you design your workplace to compensate for the fact that employees may become complacent? When you think about complacency, it's everywhere. It's kind of like the old fire triangle (see Figure 27) with the fuel, the spark or heat, and the oxygen. We all know you need to have all three of those things for a fire. But because the spark and the fuel are much easier to manage than the oxygen that's everywhere, most fire prevention activities are aimed at not starting the fire in the first place by eliminating the

Figure 27
The Fire Triangle

ignition source (Don't Play with Matches), and then by making sure that combustibles are stored away from any ignition source (DANGER—Flammable Material: Keep Fire Away).

Because oxygen is everywhere, it's more difficult to focus on, so we tend to look more at the spark and the fuel. You could think of complacency as being the oxygen. It's everywhere. In almost every accidental injury, it's either a driving factor in the foreground or a contributing factor in the background. So when designing complacency out of the workplace or reengineering the workplace for less complacency, you have to remember that reducing complacency is going to be a daunting task because it's everywhere. Keep in mind that we don't want to look at everything. We want to look for specific places where an intervention is going to have maximum results.

After discussing how inside-out thinking could help reduce rushing, frustration and fatigue, I asked Gary, "Do you have any examples of how you've used inside-out thinking to negate the effects of complacency?"

CHAPTER 10—*Using Inside-Out Thinking to Compensate for Complacency*

"In one facility I worked," Gary replied, "we had a large manufacturing process that was going to be really, really difficult to guard. You know, there's a lot of equipment out there that is hard to guard like papermaking machines with all their nip points and the rotating rollers, or steel or aluminum roll slitters that require lots of human interface. Anyway, the company ordered a new manufacturing process that consisted of several machines and pieces of equipment; the manufacturing engineers really loved the way it worked. When all the equipment arrived, the maintenance technicians set it up. When that was about finished, the engineers came to me with a question, 'So, do we need anything else to guard this?' Now, most equipment manufacturers do a fair job of guarding their machines from a letter of the law, compliance standpoint. But when one piece of equipment gets integrated into a manufacturing process with other equipment, the transfer point between machines gets tricky. It was a typical foul-up. The engineers should have had this discussion about the safety of the operator and how to guard the manufacturing process with the safety department right from the beginning, but they didn't. When I started looking at what it would take, I looked at fixed and adjustable guards, photoelectric and RF devices, interlocked gates, you name it, but everything I looked at still had problems. After I'd been at it awhile, the engineers wanted to know what I thought it would cost. 'Probably $1 million,' I said."

"Sometimes when you put guards on a process, you guard it to prevent a hazard from contacting a person, for example, part of a machine, or the piece it's working on, breaks, flies out of the machine and hits someone. However, most of the time you're guarding something that nobody in their right mind would get into. They wouldn't knowingly contact the hazard. However, sooner or later everybody becomes a little complacent. This makes it easy for them to take their mind off what they're doing and even stick their hand into a place where it doesn't belong. People also take their eyes off what they're doing, trip over something, and sometimes lose their balance and fall. When guarding is done right, its efficiency is amazing. It doesn't matter how complacent an employee might be, whether they've got their mind on the game or they're paying attention to what they're doing or not, if a guard is in place, they can't inadvertently get to the hazardous energy."

"So I used a little inside-out thinking to compensate for complacency. I thought, 'Okay, we're only going to have one person operating this manufacturing process. It's in a very remote part of the facility, so we can restrict access to this part of the building. What if we just guard the employee?' What I meant by that was, what if we put the employee in an area away from the manufacturing process where he couldn't possibly be injured? We could even put a restroom in that area for the operator's comfort. Although they'll be separated from the process, the operator will have all of the controls necessary to operate it, and we'll just guard this area. So if the employee kind of forgets, for some reason, what they're doing and goes outside the area, some type of an alarm, maybe a buzzer, will sound and give the operator a few seconds to return to the controlled area before the process will shut down. The buzzer is an outside stimulus intended to bring the operator back to thinking about what he or she is doing. The idea was to isolate the operator in this area, and any movement outside of that area required the process to be shut down automatically. So, instead of guarding the manufacturing process, we took a different tack. We guarded the *person*. In other words we prevented the employee from coming in contact with any hazardous energy, even though there was plenty around, by placing them in a work envelope that if they went out of it, then all of the hazardous energy would have to be shut down. I like to call this a 'safety envelope.'"

"This is obviously an unusual example but it I think gets the point across rather well," I responded. "You mentioned that because the complacency issue was addressed after the piece of equipment was installed, and not right from the beginning, it became a complex and expensive thing to fix, right?"

"Right," Gary responded.

"It would obviously be better not to do it that way," I continued. "Wouldn't it have been better, when considering a new process or new piece of equipment, to take the four states and the four critical errors into account during the planning stage so that you're thinking about rushing, frustration, fatigue, and complacency, and eyes not on task, mind not on task, line-of-fire, and balance, traction, grip issues right from the beginning?"

"Exactly," Gary said. "What I want manufacturing engineers to do is when they're starting to think about buying a piece of equipment or setting up a system, before they ever get to the point where they're getting a quote, they need to involve the safety people. In addition to their compliance-related skills, the safety people also need to bring with them their risk recognition skills that include the four states and the four errors. The engineers need to be trained in those risk recognition skills too."

Gary was implying that if we want to get to where we

need to be, manufacturing engineers need to understand these concepts—the states, the errors and the four critical error reduction techniques. They're going to need to appreciate the implications of the state-to-error risk pattern that's involved in over 95 percent of the injuries. They're going to need to buy into the concepts and learn them well enough so they can start approaching their job from an inside-out perspective.

"That might sound very challenging to a number of people," I said.

"Oh, I think it is," Gary said. "Their entire training has been production oriented. Everything that they've received in college, everything they've received in on-the-job training, has all been things, things, things—not people. Getting these concepts across is a challenge, but I know it can be done because I've done it. One requirement I put in place was that when manufacturing engineers were considering a new process, the safety people had to be involved at various stages of the design process—10 percent, 25 percent and 60 percent complete. Now initially, the engineers thought that this requirement was just because I wanted them thinking about legal compliance. That was true, but I also wanted to get involved with them so I could start talking to them about the states and the errors, because they could easily design those into the system without knowing it. Requiring the participation of the safety people gave them the opportunity to impact the design right from the beginning. It also provided an occasion for additional training about the concepts and mentoring the engineers throughout the design process. This approach turned the typical 'Oh great, I *have to go* to SafeStart training' into an 'Okay, now I see how those factors affect everything I do' perspective. Once you get them thinking inside out, the marvelous thing that happens is that they not only apply it to their traditional designs, but now they start thinking, 'Well, wait a second, what if we do this? Or what if we do that?' Innovation increases dramatically as they think about the states and the errors. So it can be a really,

really positive thing."

"Because the safety people were now involved at specific points in the design and procurement stages—in other words, before we're going out for quotes—they could ensure that the specifications addressed safety issues, including human factors. For instance, the specification could require that the machine manufacturer has to design the appropriate tools to do maintenance on the equipment, and those tools have to be included in the quote. Furthermore, the tools have to be stationed in a position very close to where they are needed to do the maintenance changes. Grease points cannot be within the confinement, but must extend to the edge of the fence or the guard so that the maintenance people can lubricate the equipment from outside the point of operation rather than inside it."

"When these kinds of specifications are written from an inside-out perspective and implemented by the manufacturer, the maintenance people can perform their functions without shutting down the piece of equipment, which would make everybody all upset because they're losing production time. Now they can lubricate the equipment right while the machine is running— and do it safely! This type of design increases both productivity and quality. It accounts for human factors and ensures that the design of the process does not encourage them to rush or become frustrated. They won't be pressured to get things fixed right away because they're fixing them in real time while the equipment is still operating."

"So for the engineers, Gary, I'm thinking you not only have to do enough training so they thoroughly understand the concepts, but there has to be enough of a motivational component in the training so that they actually buy into the idea of treating rushing, frustration, fatigue and complacency with the same respect that they would treat a physical hazard?"

"Rather than 'buy-in,' I'd rather use the term 'embrace,'" Gary said. "They need to embrace the process. If they embrace it, then it becomes a normal part of their job, and it doesn't matter who the safety person is or if there are changes in personnel. Once they start to embrace inside-out thinking during the design process, eventually it will be part of their culture."

"The leadership group needs to understand that probably the most important paradigm shift, in terms of reengineering the workplace from the inside out, is getting manufacturing engineers to accept or 'embrace' the idea that rushing, frustration, fatigue and complacency are every bit as real as the other physical hazards they know about, and that they have to be guarded or engineered out of the system too."

"Exactly," Gary said.

CHAPTER 11—*Administering Training and Mentoring from the Inside Out*

This chapter is about administering training and mentoring from an inside-out perspective. In an earlier chapter, we talked about ensuring that the training we're providing to employees is effective, and that the training is going to go so much better if the leadership group leads, as opposed to trying to do SafeStart training in spite of the leadership group. We also talked about making sure the leadership group realizes that they have an important role in this training, and that if somebody doesn't understand the concepts and techniques, it's their responsibility to learn it well enough to coach those people through. There are several other issues in managing the training function that we need to address: How, for example, do you ensure that people attend the training? What is the plan to train new employees when they join the organization? When should the refresher training using the *Extended Application Units* be scheduled? What about make-up sessions for people who missed the originally scheduled sessions—and many other issues. I asked Gary about this.

"One of the biggest forms of resistance we have—to any training—is the logistics," Gary said. "It's questions like, 'How do we get the people to the training?' 'Can we keep our production lines going?' 'How are we going to take care of make-ups (the people who aren't there when there are sessions going on)?' 'Why do we have to schedule *Extended Application Units*?' All of these logistical issues aren't easy. A major car manufacturer I know of took half the people off the assembly line, and they slowed the line down to half speed. When the training session was over for the first half of the employees, they got back on the floor so that the second half could attend training. Each minute of production time cost the facility $10,000."

"$11,000, as I recall," I said.

"Okay, $11,000. But here's my point: Solving all of these logistical issues requires support from senior leadership. If you leave the solution to first line managers, they're just going to tell you, 'We can't do it. It's impossible.' They'll say, 'You don't understand our system; you don't understand what we have to do here.' Believe me, after twelve years, we've been in every kind of facility that you can imagine, and there's always a way to handle the logistics. But the leadership group has to help make sure that the training happens, and that good records are kept. We have to know exactly who has attended the training and who hasn't. We have to make sure there are make-up sessions so everyone can receive the training. All of this takes support from senior leadership. But we also need their support beyond the initial training to ensure that all employees receive the *Extended Application Unit* training as well."

That reminded me of an implementation that took place in Australia for a large cement company—or, rather, it was three pilot sites for this company. I remember seeing some data that compared the total recordable injury frequency rate (equivalent to OSHA recordables) before and after SafeStart was implemented at those three sites (see Figure 28). About 95 percent of employees and contractors at all three sites completed the course. The results were quite impressive, especially when you realize that the improvement was achieved in only five months. But I've wondered about the people at the three sites who weren't able to attend the training, the ones who accounted for such a large percentage of the injuries in the second year, and about all of the other people in the organization who weren't part of the pilot. When employees miss out on the training, for whatever reason, they are not able to take advantage of the benefit of improved safety awareness and the critical error reduction techniques their co-workers learned in the SafeStart training. Those employees are more at risk than their co-workers who received the training. The company misses out not only on improved safety performance, but also gains in productivity, quality, lower operational costs and improved customer service.

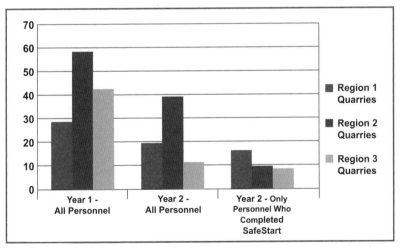

Figure 28
Cement Company Total Recordable Injury Frequency Rate

In addition to getting employees to the training sessions, it's also important to motivate them. I've talked about supervisors needing to know the concepts and techniques well enough so that they can coach their employees. It's really no different than, say, lockout/tagout. The supervisor is responsible to ensure that the employee is competent to perform the task. SafeStart training *does* put another burden upon an already burdened supervisor. But when you account for the critical error reduction techniques helping to mitigate the effects of rushing, frustration, fatigue and complacency, and when you see the positive effects on safety performance reflected in productivity and quality gains, then it ceases to be a burden and becomes a powerful tool to improve management efficiency.

While I think the primary responsibility for making sure employees understand these concepts and techniques should still remain with the supervisor, many companies have looked at relieving some of the burden by having a more formalized mentoring program.

The mentoring process will happen naturally; it just may not happen the way you want. The new employee might pair himself or herself up with an employee who doesn't always follow the rules. However, if you had a mentoring program that was formalized, where the mentor is consciously chosen for a specific employee, that mentor could make sure that the new employee was getting the training. They could talk about their experience: how much it helped them or how much it helped their family. Such a program would be much better than placing the whole burden on the supervisor.

Unfortunately, from what I've seen in the past 25 years, quite often upper management seems to think it's okay to just pass the responsibility down to the first line supervisor. They expect that the supervisor—who might have 60 to 100 employees working under him or her—to be a mentor or a coach. Yet some of these supervisors can't even get their hands around organizing the schedule for the training sessions. To expect that we're going to get a great deal of mentoring and coaching out of supervisors is probably being a bit idealistic, especially with the workload they already have. However, what we do need to understand is that as new employees come on board, they will be seeking a mentor. I remember Gary joking about this once, saying, "They at least need someone to tell them where the washrooms and break rooms are, and to give them the inside scoop about what's really going on." They will be seeking that out all right. But what we need to do is make sure the person they talk to starts talking to them about the importance of these four states and these four errors to help out the supervisor so we're not throwing the whole mentoring burden on them. I asked Gary about this, too.

"Part of my training when I got my EMBA was in organizational theory. One of the things we talked about was the concept of the 'unofficial boss.' Of course, there is always the supervisor. But, in addition to the supervisor, there is always somebody, usually another wage employee, who is looked up to by their peers. That person, frankly, is the real boss—the unofficial boss. Even if you've never put a formal mentoring process in place, you still have one whether you know it or not, because mentoring is going on, but it might be leading in the wrong direction. You might have a maintenance person saying, 'Well, I know that they told you *that* when you hired on, but we really don't have to lock it out. All we do here is shut off the power.' Or you might have an employee tell the new guy, 'Well, you're only going to be about 10 or 12 feet (3 or 3.7 meters) in the air, so if you fall, you can land on your feet and nobody will know. I mean, nobody's out here, and it's a waste of time to go all the way back and get your fall arrest equipment to do the job.' If I had a dollar for every time somebody told me, 'That's not the way we do it around here; I don't care what *they* told you,' I might have actually retired by now. So there *is* mentoring going on. What we need to do is control the mentoring and make it positive, something that's productive for the organization and the employees. Let's make sure that new employees are matched up with someone who demonstrates the value of knowing about the four states, the four errors and the four critical error reduction techniques, as well as all of the compliance issues."

The last thing I wanted to cover was the issue of what kind of training to provide external contractors? Almost every company uses external contractors to some degree or another. In some places, they almost seem like permanent employees since they're there all the time and work regular shifts. Sometimes they're doing things that the regular maintenance or operations people can't, like a major turnaround at a refinery or petrochemical plant, or replacing the turbine in a generator at a power plant. Other times they're on long-term contracts, a lot more than just six or eight weeks, and may provide all of the maintenance at a facility. In either case, they work very closely with the regular employees and, as a result, they impact what goes on. In some cases, they can directly or indirectly contribute to an injury. Many companies have decided to bite the bullet one way or another, and suggest—or, in some cases mandate—that the contractors receive and go through this training on the four states, the four errors, and the four critical error reduction techniques. But we all know there are issues with trying to get contractors who don't work for you to do stuff.

I asked Gary, "What did you see and what worked for you in getting the contractors on board?"

"In some cases, you're right, they almost seem like permanent employees because they're there all the time," Gary responded. "So why wouldn't you want to extend something that's working so well for you over to your contractors? In the long run their injuries are your expense because they're going to transfer that cost to somebody else, and that somebody else is you. Think about it, there's a lot of interaction between these contract employees and your full-time employees. When that happens, you want those contract employees to not be doing anything that would endanger your employees. The interesting thing is that from a compliance standpoint, contractors generally do a little bit better job than full-time employees. Their compliance performance may be better because their contract requires them to follow certain safety procedures. If they failed to follow the safety procedures, their contract might not be renewed. So from a compliance standpoint, they generally are doing everything about right. But if they don't receive training about the effect of human error on personal safety, they have no idea about human factors. They really need to understand rushing, frustration, fatigue and complacency—the human factors piece. If your employees have had this training, but other contract employees in the workplace haven't been exposed to these concepts, you have two distinct groups within the same organization who view safety from two completely different perspectives."

"You have exactly the same problem when one part of a company gets SafeStart training and another part does not get SafeStart training. It always results in problems with those two groups interfacing with each other because they don't have the same baseline knowledge about personal safety, and they might not be familiar with the new human factors vocabulary. So it just makes sense that you'd have your permanent contractors trained."

"The best example I've seen of the problem that not training everyone can cause was at a major manufacturer in the southern part of the United States where I did a formal audit of their entire safety system including their contractors. In this facility, contract employees did *all* the maintenance. While performing these maintenance activities, contract employees had to directly interface with permanent employees. Unfortunately, this interface was not going well. The permanent employees had been trained in SafeStart, and their safety culture had matured beyond simple compliance. They understood the human factors that lead to higher risk. In fact they embraced the SafeStart concepts and techniques so thoroughly that SafeStart became a way of life for them every day. The permanent employees felt that the contractors were creating higher risk not only for themselves but also for the full-time employees. This wasn't because the contractors were violating the safety regulations. It was because they didn't know about the human factors that increased risk and what they could do to minimize their effect. It wasn't until the contractors went through the SafeStart training that they came to understand the issues the permanent employees had with them."

"So you really need to be thinking about training anybody who works in the facility on a regular basis. But it isn't just your typical contractors: It includes the telephone technicians who troubleshoot the phone system, the specialists who work on the data network infrastructure, or the people who install and maintain the fire alarm and security systems. Anybody that comes through your door on a regular basis ought to have SafeStart training in addition to any required compliance training."

Gary's story about the major manufacturer got me thinking about a forest products company for which I did a lot of work. I can remember there was an issue during the winter when it was slippery on the top of the lumber trucks. They had people falling 14 feet (4.3 meters) from the top of the trucks to the ground. And as you can easily imagine, some of those people didn't just get up and scamper away after they had the wind knocked out of them. The good news was that none of the people who fell off the trucks were their employees; they were all contractors.

Now I suppose you could say that the company didn't have to address this because the contractors' injuries weren't their responsibility. After all, the contractors were the only ones allowed on top of the truck, because it was their own truck. However, the company recognized that they had a responsibility to provide a convenient anchor point for fall protection so that if the truckers lost their balance, they didn't get severely hurt. So, even though the contractors were the only ones who needed to wear fall arrest equipment, they made an investment in a reasonably accessible anchor point and built a structure with concrete foundations and steel pipes going out across the loading area so there were places to attach a lanyard. They knew they couldn't stop the loss of balance, traction or grip in the first place, but they did foot the bill to pay for the engineering and construction.

I related this story to Gary and then asked him, "Do you think that was enough for the truckers to start to wear a fall arrest harness and use the convenient, accessible anchor point?"

"My guess is no," Gary answered. "Unless they were instructed or ordered to use the anchor point, I doubt if it would ever happen. I also think the company would have been a little bit hesitant to tell the truckers to do that because of the issue of transfer of liability. So what was the end result?"

"I've heard you mention the liability issue many times before, Gary," I responded. "That's the problem with enforcement: it's a resource-intensive activity. In other words, now you have to have somebody going out and checking. Then if things aren't right, you have to have some way of dealing with the lack of compliance. When the company looked at their options, it wasn't as if they were in an area where they had an unlimited supply of contractors to choose from. They were also limited in their own resources. They couldn't just give the yard superintendent full-time fall arrest harness inspection duty because he already had a lot of other things to do. They *did* contemplate hiring a supervisor who could enforce the use of fall arrest harnesses, but that would have cost them $100,000 the first year, $100,000 the second year, and on and on. It was going to be a huge expense, half a million bucks over five years. So you can certainly see how, if you were looking at this strictly from a return on investment (ROI) point of view, you're not getting anywhere very quickly."

"The company decided that it would probably be more cost effective to take a shot at doing something a little different. Now, you have to appreciate that this was an unproven idea for them back then. What they decided to do was pay for me to do a training session for the contractors on the four states, the four errors and the four critical error reduction techniques. They wanted me to explain why the company had installed all of the anchor points, and to encourage the truckers to wear a fall arrest harness and attach it to an anchor point. The whole purpose of the training was to get the truckers to comply voluntarily. They hoped that the superintendent would only have to spot check compliance occasionally, so they could avoid paying $100,000 a year for another supervisor. They looked at that as being a much more cost-effective solution. So, *all* I had to do was motivate the truckers, who hadn't used a fall arrest system for twenty years."

"And didn't want to use it," Gary added.

"True, they didn't want to use the fall arrest system," I agreed. "But they weren't that bad, after all, they were just people. When they understood the four states and the four errors, and the significance of the 'self' area, there wasn't a whole lot of arguing when they began to look at their own injuries. So this forest products company improved what they were doing with a combination of redesigning the workplace and training. They eventually got the contractors on board, and that improved the company's credibility with their own workforce. I mean, they were telling their workforce that they wanted to be the safest forest products company in the country. But their workers were saying, 'You aren't even making the truckers wear fall arrest harnesses on top of the trucks, and we've had two of them fall off this winter.' In the end, it was all about putting your money where your mouth is. While the company did spend *some* money, they didn't want to put more money where their mouth was by paying for a supervisor to enforce the rule. This is one of the issues everybody in safety has to come to grips with: If your solution is enforcement, then it is not a cheap, long-term solution. That's the problem with enforcement. However, once you get things happening peer to peer in real time, as in your Stage 5 Gary, it's free. Police are not free. That's something you need to consider from an administrative point of view, in terms of inside-out thinking."

CHAPTER 12—*Changing Your Safety Culture*

This chapter is really about trying to tie together all of the elements Gary and I have talked about thus far. We've talked about the culture you need to create: an inside-out culture that includes both human factors and critical errors, a culture much more balanced than the outside-in, extrinsic control of the traditional safety management system. But how do you get from here to there?

"What steps do safety professionals need to be aware of as they go through this transformation from an outside in culture to inside out one?" I asked.

"Okay, but first let me talk a little bit about culture from a safety management perspective." Gary replied. "Culture has probably been one of the most overused words in the last twenty years. Everybody's trying to create a culture, and in my business, they've been trying to create a safety culture. What I want to point out is that while we want to have a safety culture, and we want to have it from the inside out, culture is what comes after the equal sign. Let me explain. Here's the formula: X+Y+Z = C, where 'X,' 'Y' and 'Z' are some of the variables that impact culture, and 'C,' of course stands for culture. Culture isn't the X, Y and Z. Culture is what it is. Sure, it's the result of all your efforts, but you can't create a culture. You certainly can't control the culture. What you can do is control the things that are on the other side of the equal sign, and I call those things your climate. Now, I'm not saying that a positive safety culture is bad, because we all want to get there. But we can't control culture. However, we can control climate, the Xs, Ys and Zs, those things that are in front of the equal sign. We've got a chance of controlling those, and when we do change the climate, eventually our culture changes."

Hearing this, I remembered an article with a title something like "Creating the Safety Culture" I read in about 1991. Even back then, the article reminded me a bit of the Aesop's fable called "Belling the Cat." What I recall from the early '90s was that there was way, way more talk about how good it would be if the bell were on the cat—if we had this positive safety culture. There wasn't nearly enough, at least from my perspective, on how we could get that culture. I thought they really did a great job of overselling the benefits of a positive safety culture to safety professionals who were already fairly well aware of the benefits, but they were less than helpful in describing the path of how to get there. Although I don't think I've ever said it as well as Gary does in terms of "Culture is what comes after the equal sign," you know, it's not the X plus Y plus Z.

I asked Gary to tell us what he thought the X, Y and Z were. In other words, what are the critical leverage points we *can* influence to eventually impact our culture.

"You remember the diagram that showed the five stages to world-class safety, don't you?" Gary asked."

"We looked at that a few chapters back," I replied.

"Okay, well, back then when I developed the diagram," Gary continued, "I believed along with a lot of other people that when a company went through these five stages their culture changed (see Figure 29). People may have differed on the number of steps, but it was essentially a linear process: as you progressed from stage to stage your culture changed. So let me briefly review what those stages involved. When you look at the diamond, you'll notice that it is divided into five areas that differ in size. The size of the area represents the number of organizations in each stage. For example, there are more companies in Stage 2 than Stage 1."

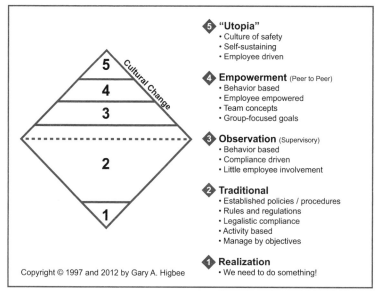

Figure 29
Five Stages to World-Class Safety

Realization—Stage 1:

"The type of organization you will find here is one that is not performing well. High worker injury and severity rates, potential litigation, along with regulator citations are all part of this organization's climate. A company will remain in Stage 1 until they realize that things have to change and at least keep up with the safety regulations."

Traditional—Stage 2:

"Stage 2 is about all the rules, policies, procedures and regulations we are required to comply with, and hopefully we do it well. On the diagram near the top of Stage 2 there's a dotted blue line to indicate the point at which the organization is basically meeting their compliance requirements. The space above the dotted blue line is used to indicate that employee involvement plays a very important part in the safety process. A company at the top of Stage 2 has a comprehensive safety management system in place. They have achieved a level of compliance where they are meeting all of their legal obligations, and have the active involvement of employees in the process. Frankly, this is the point most organizations are trying to reach. Unfortunately, a company's safety performance at Stage 2 might be legally acceptable, but it's far from where it needs to be. I talked a little bit before about my friend being killed in a facility with excellent safety performance as far as complying with the regulations went, and how I was really frustrated with this situation. Although none of the regulations required a company to do anything to go beyond Stage 2, I just knew there had to be something more because even at the top of Stage 2, when we were meeting all of our legal compliance requirements, we were still having slips, trips and falls, back injuries, sprains and strains, cuts, contusions and abrasions."

Observation—Stage 3:

"This was primarily a supervisor to employee observation process looking for intentional unsafe activity. This process seemed to be one way to look for and discover gaps in the safety system. The observations were built on top of the traditional safety systems found in Stage 2, so it was a step forward. With proper positive reinforcement for safe behavior and positive correction for at-risk behavior, this stage didn't do everything that was needed, but it did help. The observation process did involve issues with discipline for intentional at-risk activity, but the biggest issue for me was that it didn't seem to help with unintentional activity, the mistakes employees never intended to make in the first place. So

I added the observation process to what I was doing."

Empowerment—Stage 4:

"Then after a while I thought, well, since it's working well for supervisors to observe employees, why don't we get the employees doing observations themselves? This was hardly an original idea. When I benchmarked other companies, I found a number of organizations already doing peer-to-peer observations. Stage 4 gets employees talking to each other about safety and increases overall awareness. However, these peer-to-peer observations were formal, regularly scheduled encounters."

"Utopia"—Stage 5:

"Do you remember the story I mentioned about people removing asbestos from a power plant and I said I thought we'd..."

"Reached 'Utopia'"? I interrupted.

"Yes, we reached Stage 5 where peer-to-peer observations were happening in real time, not formally scheduled observations. I thought we had reached the ultimate where we've just got each other's back. So those were the stages we went through. That was the X plus Y plus Z, which got us to what I used to think was 'Utopia.'"

That was the process I also used to think you had to go through in order to get a safety culture where peer-to-peer interventions occurred in real time. And although I didn't have a Stage 1 and a Stage 2 like Gary did, I accepted the idea that you couldn't move to a peer-to-peer observation process unless you were essentially in compliance with the regulations first. So when I would talk to senior managers about what type of behavior-based safety process they should be looking at and who should be making observations, I'd say, "I know you want to move right into a peer-to-peer observation process because that's what everybody's talking about these days, but you haven't actually done the work required for this stage. If you tried to do it, you could have an employee going up to another employee after observing them on a grinding wheel and asking, 'What about the face shield? Shouldn't you protect your face? You've got your eyes protected with the safety glasses, but what about your face?' And the employee at the grinding wheel might say, 'Well, my supervisor doesn't even make me wear the face shield, so I don't know where you're coming from but quite frankly you can go back there.' In other words, it would be very difficult from a peer-to-peer standpoint to get employees to enforce rules that supervisors ignored or to use positive correction for at-risk behaviors that supervisors condoned.

Managers accepted this advice without question. So I also believed that this cultural change was a series of steps or stages, and that each one would take a certain amount of time depending on how much effort the company put into it. For example, if you didn't put enough effort in Stage 3, you might never get to Stage 4. On the other hand, if you exerted lots of effort in Stage 3, you might get to Stage 4 more quickly. And if you gave people enough support in Stage 4, then you would eventually get to Stage 5. However, if the leadership groups didn't support the peer-to-peer observation process, you might never get to Stage 5 or, even worse, your observation process might fail. You see, I completely accepted the idea that changing your safety culture involved a slow linear transformation from Stage 3 to Stage 4 to Stage 5, and the amount of time it took to go from stage to stage depended upon how committed you were and how much effort you put into it.

Then something happened to change my thinking completely. It was right at the beginning of 1999 at a pipe coating plant in a town in southern Louisiana whose name I can't remember right now. The unofficial rumor was that somewhere around 30 percent of the employees were on parole from a federal prison—not a state prison—a federal prison. From a safety engineering point of view, the plant was a disaster. Anyone could see that they didn't have a safety person. I mean, the machine noise was deafening. It had to be somewhere around 120 to 130 decibels, and hardly anybody was wearing hearing protection. When I considered how long it would have taken to get them to Stage 5 using a traditional behavior-based safety process, quite frankly, I would have been very hesitant to say *when* or *if* they would ever get there. The company lacked any real leadership, and didn't have a safety management system or adequate safety engineering. They were barely above the line separating Stage 1 from Stage 2. Nevertheless, they implemented SafeStart, and then that company of 360 employees went the next year without a recordable injury. That rocked my boat, pleasantly of course, but it wasn't what I expected.

After I recounted the story to Gary, I asked him whether he had experienced anything like it.

"I had the same thing happen to me," Gary replied. "I bought into this Stage 1, 2, 3, 4, 5 theory, but..."

"Well, it *is* your theory, isn't it?" I interrupted.

"Yeah, it's my theory," Gary continued. "But the thing that surprised me was that it almost didn't matter what stage the company was in. Quite a few companies would tell me that they were in really good shape, but when I got there and had a chance to look around, they hadn't even met the minimal requirements for Stage 2. Some had a rudimentary observation process in place, but it wasn't functioning very well at all. Now they wanted *me* to help them implement SafeStart. I couldn't stop thinking, 'This will *never* work!' I mean, I can't do anything until they clean up this mess. They've got to get a whole lot better with compliance before SafeStart stands a chance. But the fact of the matter was, it worked—SafeStart worked. In fact, here's what was happening: The SafeStart training for whatever reason—I think it's because it's so intuitive and easy to understand—was actually helping the process of getting the Stage 2 problems under control. In other words, because of SafeStart, employees were now seeing things that were hazardous, where before they just accepted the hazards as a normal part of the operation. It also affected their observation process, which was not very good or efficient. SafeStart gave the observers something very different to talk about: rushing, frustration, fatigue and complacency, and the role they play in human error. Because the employees understood the state to error risk pattern and were able to use the critical error reduction techniques every day, the company made a big leap in terms of changing their safety culture. That taught me that a company didn't need to get to 'Utopia' before they could implement SafeStart. In fact, SafeStart proved to be effective no matter what stage they were at. It actually improved the traditional safety program and the observation and feedback process. It's almost like SafeStart pulled the whole safety management system upward. It acted just like a catalyst to improve their safety program and catapult it to a higher level."

I knew this was true. It had happened in a lot of companies where our consultants or I had worked, places where they were not anywhere close to the level they had said they were when we talked to them before we arrived. Back in the old days before SafeStart, I would have taken my behavior-based safety consultant's hat off and put on my safety engineer's hat. I didn't know the technical aspects as well as Gary, but I knew enough to get them to start working on the basics. They needed to start looking at machine guarding, PPE, lighting, ventilation, basic 101 stuff—they *had* to start there. The reason was simple: In the old behavior-based safety days, people would be more than willing to come forward with these unsafe conditions when they were being observed. When I would try to talk to them about their at-risk behavior, they'd point out, "What about the equipment over there, where the guard is broken?" or "What about this switch that doesn't work?" These are important issues, but this isn't what an observation should be all about. These things should already have been taken care of before now. So I'd have to start working with them to shore up their basic safety program since they still had miles to go before they were ready for Stage 3. However after SafeStart, when we started to talk to employees about the "self" area and the three sources of unexpected, the four critical errors and the four states, and the state-to-error risk pattern, things changed so quickly that within weeks, companies were actually getting some peer-to-peer intervention in real time.

And I'm not talking about places that already had positive safety cultures. I'm talking about places like railroads where they had gone down the outside-in road for so long that every time somebody got hurt, the company made a new rule. And when somebody *else* got hurt, they made another new rule, until eventually, they ended up with 743 safety rules! Accident investigation at this place was simple: find out what happened, and then make sure you had the right number of the rule that was broken—end of investigation, case closed. The employee broke rule number 438, and according to the rule book, here's the discipline that's associated with breaking rule number 438. Everybody knew that if you followed all of these 743 rules, your railroad could never compete; everything would be too slow. And yet when SafeStart was introduced, we saw peer-to-peer intervention out there in the yard within three months. Train crews would say things like, "Hey, watch line-of-fire" or "Hey, eyes on task." It started off sort of like a joke. I remember Dave Moorehead, the union rep, saying it started off with guys kind of kidding each other, saying, "Look out, line-of-fire." But when the situation involved the potential for imminent injury, people became serious and said, "Hey, eyes on task," like they really meant it. Pretty soon the trainmen were using SafeStart concepts in their vocabulary in the shop and out in the yard. The railroad was getting peer-to-peer intervention in real time, and it happened within months. They were also getting injury reduction. One mechanical division with 300 employees went 9 months without any injuries.

Gary and I don't always get the chance to work together on the same SafeStart implementation, like we did for the U.S. Postal Service. I worked on the railroad job by myself, but when we did get together, I would give him an update on what was going on, and he would tell me what he'd been doing. I asked Gary whether he remembered that conversation we had years ago when I told him about the mechanical division that went to peer-to-peer within three months and went nine months without so much as a first aid.

"I remember that conversation very well," Gary said. "We both kind of looked at each other, and you said, 'Gary, you *won't* believe what I'm seeing out there.' And I said, 'Really? Because what I'm seeing is really amazing too.' We were both so surprised, because we didn't expect SafeStart to work until the compliance issues were taken care of and an observation-feedback process was established. In fact, I thought all along that our observation program SafeTrack® would be our bellwether product, and then SafeStart would be the product that would give a company a little extra boost to get to Stage 5. Back when SafeStart began, I didn't think we'd be using it much because of how long it would take most companies to achieve 'Utopia.' Looking back I couldn't have been more wrong. We both saw significant changes quickly, and the concepts and techniques didn't offend people. In fact, the observation process sometimes was more offensive than the employees' realization of the importance of the 'self' area. So as companies implemented SafeStart, what they were really doing is changing their safety culture from the inside out. Like I said, it was amazing."

"After we both realized that a company didn't have to go through this linear, fairly time-consuming process to get from Stage 2 to 3 to 4 and eventually to Stage 5," I said, "we stumbled upon another way SafeStart could help change the safety culture. When we worked with the U.S. Postal Service in Denver, we ran into an incredible degree of apathy. Their employees, like everybody else, were motivated not to get hurt. The problem was that they worked in a very low-risk workplace. The likelihood that anybody could get killed in that workplace wasn't very high because there weren't a lot of mobile equipment or hazardous high-energy processes going on—employees were just moving mail. Hardly any of the employees saw the value of SafeStart until we started talking about 'the family.' Then this apathetic workforce got engaged in a big way. You see, their employees didn't find themselves personally vulnerable from a safety point of view. They all thought they were safe enough already. But when we mentioned the family, we discovered something that really concerned them, something they felt vulnerable about. Did they think their family, their teenagers, were safe enough already? No, they didn't, not at all. When they recognized that SafeStart was something their families really needed, they became engaged because they cared about their families. In the process of trying to create an inside-out culture at work, we learned that the easiest way to do that was to involve the family. Once we realized how powerful this motivator was, we started pushing on the family button—hard."

"I knew that a lot of companies had dabbled with the idea of trying to get the family involved. For years and years, many safety professionals had recognized what a motivational component the family is. I'd hear them say things like, 'We want them to go home safe to their families.' They would remind employees, 'If you get hurt here at work, how are you going to be able to play with your boys or throw them the football? If you were blind, you wouldn't be able to even see them.' They knew how motivating the family was to their workers, as it is to all of us. They knew the family card was important, but they weren't playing it properly. They didn't understand that their employees weren't worried about their own personal safety; they were worried about their families' safety."

"Companies have been trying to make inroads into improving off-the-job safety for a long time. But there's only so much you can do in terms of family days, picnics, and passing out flashlights, first aid kits and fire blankets. They probably have had somebody from the fire department come in and talk to everybody about having an escape route, portable escape ladders available from upstairs windows, fire blankets, how to use a fire extinguisher and what kind of fire extinguisher to use. And they might have had a police officer come in and talk about parking in a well-lit spot, locking your car up, not leaving valuables visible, and locking your doors and windows at home. However, the leading cause of death for ages 1 through 44 is unintentional, or accidental, injury. It's not violent crime, and it's not disease. Fire, flame and smoke account for only 2.5 percent. But falls account for 20 percent, and motor vehicle accidents account for another 28 percent. [5] The police and fire don't normally talk about that. They don't get into the hitting your head; they're not getting into the cuts and contusions, the sprains and strains, and all of the typical day in, day out injuries."

"You know, if we hadn't had that experience with the post office, I wonder how long it would have taken us to see the real significance of involving the family. That started us on the path to make *SafeStart Home* more robust by including something for every member of the family, adding more DVDs and revamping

---

[5]National Safety Council, *Injury Facts, 2011 Edition*, Itasca, IL, pp. 8–11. The data is from 2007.

the online course. When companies started using *SafeStart Home*, they found that it helped to change their culture from one of control and compliance, to a positive safety culture that reflected the company's care and concern for employees and their families."

"It is amazing that what happened at the post office kind of started us down that path," Gary said. "Now we're taking SafeStart even further. We now tell companies that if they want to foster a positive safety culture at work that reflects care and concern, they should take it past the family, all the way into the community."

"Many of the companies we've worked with have actually done that," I said. "They've taken it to their local high schools, middle schools, church groups, youth groups, Boy Scouts, Girl Guides, YMCA, among others. Taking these concepts and techniques to the community can really increase your credibility with your workforce in terms of getting them to believe you really care. And the care and concern you show to the family and community only increases the positive safety culture at work."

Gary nodded. "I agree that the leadership group's decision to take SafeStart to the community increases the group's credibility. But it's equally important that each member of the group personally goes first. They need to take it to their families so they can talk about it when they come back and say, 'Hey, this worked for me; this worked for my family. I've got a seven-year-old and an 11-year-old, and here's what we did. We sat down, watched the DVD, and talked about the four states and the four errors.' That really improves their credibility and their ability to relate. You see, it's hard sometimes for some of the leadership group to talk about safety when they've never worked a day in the facility, never worked a day on the line or never worked a day at the construction site."

"You *really* want to say they never worked a day in their life, don't you?" I asked.

"Well, it's tempting, isn't it?" Gary replied. "But seriously, because they aren't involved in the same activities as their workers, they can't really talk about lockout/tagout or confined space entry. But they can talk with credibility about what has worked for their family, their children, and their spouse or significant other."

I knew exactly what he was talking about. We both saw this cultural change just recently, about half a year ago, when we worked on a job with one of the safest companies in the world. Years ago they had had a very robust safety culture that displayed the company's concern for its employees. Back then, the leadership group cared about each employee; they cared about their individual safety. They knew each recordable injury was much more than a number on a spreadsheet—it was pain, it was suffering, it was needless and shouldn't have happened. They really wanted to make sure it would be prevented in the future. But then, like a lot of other major Fortune 500 companies, they faced the challenges of going global, downsizing and the fluctuating profitability of various product lines. The company changed, restructured, and reorganized. The new senior management was focused on maximizing shareholder value. Slowly, over time, the company lost their positive safety culture. But because their safety record was still very important to management, they adopted a heavy-handed, outside-in approach to safety. What little care and concern that was left in the culture quickly died out. When we got there, we found very little deficient in terms of their safety management system, or their safety and production engineering, but their culture was much, much different than what it had been before. When we started implementing these concepts and techniques, and I started talking about the SafeStart materials available for the family, long-time employees who attended the session began saying, "This is the way we used to treat people here. This is the way we used to be here." and "Now I'm finally seeing a genuine attitude of care and concern coming back. I'm seeing sincere concern for me and concern for my family."

I asked Gary whether he heard that at the sessions he conducted.

"Yeah, I heard a lot of the same sentiment. In fact, early on I saw something I didn't expect. We offered to give employees some videos to take out into the community, if they were interested. So in the first sessions, they were not only commenting on the course, 'Wow, this is something cool, we missed this. We haven't done this,' but they were also interested in the *SafeStart Home* DVDs, 'Hey, I've got a Boy Scout troop I work in, can I have some videos?' and 'I've got an Awana group or a church youth group that would like doing this stuff.' The requests came in right away. The employees were excited. It's like they were saying, 'All right, I understand this; it makes a lot of sense. Let's not just keep this within the company. Let's share it with a lot of other people too.' It was only the first session, but yet the concepts and techniques made sense to them. It was fun to watch the lights go on, the whistles go off and the attitudes change right before your eyes."

"I remember one man, I think his name was Dean," I said, "who told me, 'I've been here 42 years, and this is the best thing I have seen since I've been here.' It boggled my mind that somebody could be at the same workplace for 42 years in the first place. But later when I told the safety person what happened, he just looked at me and said, 'That person hasn't said a word in ten years. When he volunteered to be a trainer for the Trainer Certification course we were all surprised. But then to actually hear him say what he said—this is just huge.' I found out that Dean had a lot of influence in his department, and that he was respected by all his co-workers. So when this caliber of employee thought that SafeStart was the way to turn their culture around, the safety people and the leadership group were very encouraged. The constant pressure from corporate to get their injury rate down had led managers to adopt a heavy-handed approach to safety. It didn't work at all and only sent employee morale into a nose dive. But SafeStart proved to be the turning point they sought for. Dean, the guy with 42 years' experience said it best: 'Finally, a tool to help us prevent the injuries, instead of another tool to help us analyze where we went wrong.' So the concepts and techniques combined with the material for the home and off the job really turned the company and their culture around quickly."

"Another problem we've run into is with managers who continually try to scare their employees into working safer. This doesn't work. I never really became quite as aware of that as when we were doing a Trainer Certification session in Denver, and one of the participants got up and talked about his two tours of duty in Vietnam. He said, 'One of the problems with being a short Hispanic person is you're the same size as the Viet Cong, and that means you can fit into the tunnels. I was what they called a *tunnel rat*. They tie a rope to your feet, and you had to crawl through the tunnels looking for the enemy. There are snakes, poisonous snakes, in these tunnels. I hate snakes, but you can't have a light because then the enemy could see the light. You can't see the snakes, and you can't see the enemy while you're crawling through these tunnels. They tie a rope to your feet, but that's not to pull you out; that's just to measure where you died so they can send a mortar in.' He told us he did two tours of duty as a tunnel rat. After hearing this story, I remember thinking, 'And you were going to try to scare this guy with a fork truck?'

> Can you really scare grown men into being safer? And if you can, for how long can you do it? I started thinking about it from a personal perspective. I remember seeing a fatal car accident on the side of the road on I-64 outside of Richmond. I could tell there was a fatality because the paramedics weren't rushing around, and the front right wheel of the transport truck was parked over the driver's

seat of the TransAm convertible. I just knew that there was no way the driver got out alive. Traffic was backed up for miles in either direction.

"Once the rubbernecking was over, how long do you think it took before everybody was driving back to 'normal,'" I asked Gary, "you know, exceeding the speed limit, following way too close, that kind of thing?"

"My guess is maybe a couple of miles," Gary said. "Then they're right back to what they were doing before. They might have gone a few more miles, but what a lot of them are probably thinking is 'that was somebody else' and 'it will never happen to me.'"

"I don't think it was quite two miles (3 kilometers) actually," I said. "It may have taken a little more than a minute. And what we saw wasn't a video with fake special effects or fake scars either. This was the real McCoy, something we all saw with our own eyes. How long did it motivate us? Not very long. So, you can't really scare adults into being safer, at least not for very long. It just doesn't work. But you can take the orneriest, toughest, rough-and-tumble lumberjack or pipe-liner out there and ask, 'Do you worry about your kids?' and it will stop them dead in their tracks. They all worry about their kids and their families. So in terms of creating an inside-out culture, you need to move your attention or focus from your workers to your workers' families. Ultimately, you need to actually move attention or focus as best you can into the community. When you do, your employees will then believe that your care and concern is sincere. The other thing that happens as a byproduct of sharing SafeStart with the community is you get a reputation for being a desirable employer. People want to go work at your facility. They've heard good things about it and all the positive things happening there. But it all starts in the workplace. The culture begins to change when inside-out thinking permeates every aspect of the organization. It gains momentum as employees take SafeStart home to their families, and hits its stride as the concepts and techniques move into the community. Each step takes us closer and closer to the positive safety culture we need to achieve the injury reduction we want. Each step also enhances the company's reputation in the community."

"And you don't have to go through the whole linear five-stage transformation. That was the big epiphany for me. You can start creating a positive safety culture with peer-to-peer intervention in real time without having every one of your compliance ducks in a row. If you extend the SafeStart concepts and techniques to the family and then extend it to the community, you will have a positive safety culture in a much quicker period of time than what I used to think. I like what you said earlier, Gary, about SafeStart being a catalyst (see Figure 30)."

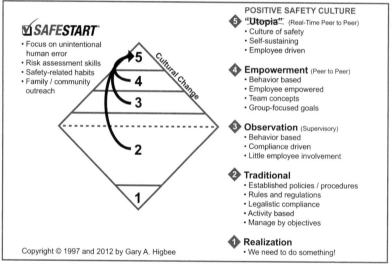

Figure 30
SafeStart: A Catalyst for World-Class Safety

"I sort of liked that too," Gary replied. "And it's true: No matter what stage an organization finds itself at, SafeStart can help them actually achieve a positive safety culture. I used to think "Utopia" was almost unattainable and that only a select few would ever get there, but not anymore. Now almost any company can have a positive safety culture in less than two years. That doesn't mean that it won't take any effort, but it is possible. We've seen it time and time again."

CHAPTER 13—*Improving Morale*

This chapter is about employee morale and how to improve it. Some people who attend one of our workshops or a conference where Gary and I or one of our consultants are speaking become concerned, especially at the beginning when we present the three sources of unexpected (see Figure 31). We use that diagram to illustrate just how high a percentage of injuries occur in the "self" area, both on and off the job. The people who are concerned worry that this message simply blames the employee, and it reminds them of a time 30 or 40 years in the past when employees got blamed for everything. In answering their concern I would try to explain that if we went back to saying "Try to be more careful" or "Try to pay more attention next time," then we *would* be going back in time. Back 30 or 40 years ago, before systems like preventive maintenance (PM) and total productive maintenance (TPM), the "self" area probably wouldn't have represented over 95 percent of our injuries. It probably would have been lower, say about 80 to 90 percent, because our engineering and equipment reliability wasn't as good.

I would also try to explain that you could look at the

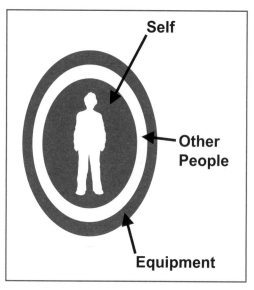

Figure 31
Three Sources of Unexpected

sources of unexpected diagram from a negative point of view and say, "Okay, you caused 97 to 99 percent of your own injuries: your error was the triggering event that started the chain reaction that led to the injury. It's your fault, you're to blame. It's your problem." You *could* view it from a negative perspective, but you *could also* look at it like this: There are a lot of things in life that I can't control. I can't control how the bridges were designed and built on the interstate highway system years ago. One bridge over the Mississippi River in Minneapolis collapsed, killing 13 people.[6] You can't control that. That's not something that I can control as a driver. When I look at the other guy hitting me, I can't control the other guy. The best I can do is stay out of his or her way, or stay out of the line-of-fire. I can control a great deal of what I am doing, but I can't control it all, which is one of the

---

[6]Matthew L. Wald, "Faulty Design Led to Minnesota Bridge Collapse, Inquiry Finds," *The New York Times on the Web* (2008, January 15), <http://www.nytimes.com/2008/01/15/washington/15bridge.html>,accessed April 9, 2012

problems. People make mistakes. They're not always doing things deliberately. Some of the things they do are unintentional, and some of those unintentional errors can get you hurt if those errors put you into contact with hazardous energy.

What I try to explain about the sources of unexpected diagram is that this is *not* a message of blame—this is an empowering message: You can't control the engineering or the environment in many cases, and you can't control the other guy, but you can control what *you* are doing. You can control your decisions, and you can control your error—not 100 percent—but the critical error reduction techniques will help you control the errors that you never meant to make in the first place, which means that you're going to be doing a better job. It's unrealistic to think that you're *ever* trying to make any mistakes. I mean, by definition a mistake is something you weren't intending to do. The critical errors you make can get you hurt, but they can also affect quality, production, maintenance or productivity.

Everybody gets frustrated when they make mistakes. Obviously, I don't know whether you (the reader) get more frustrated with yourself when you make mistakes or get more frustrated when other people make mistakes. But everybody gets frustrated when mistakes happen. So reducing the errors that people make is also going to reduce frustration, which will likely have a positive effect on safety and morale. And quite often, what makes you frustrated is that you can't get something done or you can't get something done on time. For example, you're in a rush to get somewhere, but there's construction on the highway, and it's slowing traffic down. You try to go somewhere else, but that traffic is jammed up too, so you get into this cycle of rushing and frustration. You're trying to rush but can't, so that makes you more frustrated, which makes you more inclined to rush, which makes you more frustrated when you can't. It becomes a vicious cycle. Everybody sees that vicious cycle. Push it far enough and you end up with things such as road rage. But you *can* reverse that cycle by reducing the frustration, reducing the rushing and reducing the errors. You probably have heard this saying by John Wooden, a former basketball coach at UCLA, "If you don't have time to do it right, when will you have time to do it over?"[7] The answer, of course, is that nobody has the time to do it over. So the implicit message is don't rush. Anything we can do to empower employees and help them to reduce rushing and frustration will improve morale. Making mistakes and feeling powerless rarely improves morale.

On the other hand, feeling that you're in control, feeling you have some power that you can exert, that you can do a better job of controlling some of the mistakes you never meant to make in the first place—this is empowering and will improve morale. The sources of unexpected diagram just has to be positioned properly so that people don't think you're trying to blame them for making mistakes when you're really just trying to give them tools that will help them to avoid injury in almost every situation. When you reduce errors, you reduce frustration. In some cases, you also reduce the need to rush, breaking the vicious rushing-frustration cycle. So these things are going to improve morale from a personal point of view.

In the last chapter we talked about how important it is to involve the family when you're trying to create an inside-out culture because it's an area of concern, or a point of vulnerability, for each employee. Seeing that your kids and your family are benefitting from the *SafeStart Home* DVDs you've brought home for them is obviously going

[7]John Wooden, "Quotes," *GoodReads*, <http://www.goodreads.com/author/quotes/23041 .John_Wooden>, accessed April 9, 2012

to improve morale. What could be more satisfying to a parent than seeing their kids learn things that will help them to keep from getting hurt, or hearing your kids say, "I want to watch the DVD again"? I can't tell you how many times I've had parents tell me, "They make me play it in the minivan when I'm driving now."

I wondered what Gary had seen in terms of improved morale at the places where he had implemented SafeStart, so I asked whether he had any more examples besides the ones we'd already talked about.

"Obviously, if SafeStart is implemented the right way, employees are empowered," Gary said. "Not only are they empowered, but they're empowered in a way that makes them want to do other things too. In addition to the improved safety, which is a given, the improved morale will lead to increased cooperation and efforts to improve quality, productivity and reduce cost. The post office is probably the best example I can think of."

"It's certainly the most extreme," I said.

"Absolutely the most extreme!" Gary agreed. "I don't know whether you were even able to convey how bad that situation really was. But if you wanted to find an organization that had poor morale, the post office would have been a good place to go. It was as if the poor morale was cast in concrete. It wasn't until we started talking about what we had for the family that the employees started to loosen up, get a little excited and participate in class. As they got more excited, they started to get more involved in the actual SafeStart training. When they got the DVDs to take home, their enthusiasm increased because they had something valuable to take home to their families. By the end of the course, some of the employees, who had been fairly disgruntled at the start, saw enough value in the training that they now wanted to also be trainers."

At the closing leadership session at the U.S. Postal

Service, the manager of the Denver General Mail Facility thanked both Gary and I for what we had done. But he didn't thank us for the injury reductions. I remember what he said to his group of supervisors and managers: "Guys, if anybody had told us they could come in here and within a month and a half turn the attitude of this place around so that it was positive versus negative, we all would have laughed. We would have laughed, and we would have said it's impossible. You will never turn the attitude of this place around." I remember him looking at both of us saying, "I just want you guys to know you have done this. You have done what none of us believed anyone could do. You have turned not just the safety culture around, but you have turned the culture of this place around into something positive and friendly." Both of us got a spontaneous round of applause from those supervisors and managers. It was one of the most rewarding experiences of my career up to then. But I've seen the same thing happen in more than just that postal facility. I've seen it happen at forest products companies, a steel mill, in manufacturing operations, logging camps and at oil companies. I even saw a positive turnaround at the place where the refinery manager had threatened all of the employees when he said, "You will not have a recordable injury here. And if you do, I guarantee you won't have more than one." His not too subtle implication: you won't be working here anymore. We even saw an incredibly positive turnaround at that oil company. Although it almost goes without saying, there are a lot of things that improved morale will do for your company. You're going to have a more productive, efficient operation. You're certainly going to get more employee suggestions, and you're going to have less turnover.

I asked Gary to talk about some of the less sensational indicators that SafeStart has improved, in addition to just looking at injury reductions.

department. It would be a good idea to chart them. In fact, it would be a good idea to get a handle on those costs before you even begin to implement SafeStart. Here's what you'll see: If you're making fewer mistakes, not only does your quality improve because you're making less bad product, no matter what the product is, but you're also not damaging the equipment as much, and you're going to have less downtime. So those things have a great deal of value. In fact, sometimes the cost savings from improved quality and decreased downtime from accidental equipment damage far exceeds the value of reduced injuries."

"We saw this very thing," I said, "when we worked with an auto manufacturer here in North America. When they looked at the four states, they could see that a high percentage of their quality problems were caused by rushing, frustration, fatigue and complacency. Now, politically, it's never a good idea to tell your employees that anything is more important than their safety. It's not a good idea to say that quality is more important than safety. But kudos to the Japanese auto manufacturers and other companies like them who have been able to create a quality culture where employees actively make suggestions for improvement. They chart metrics like how many employee suggestions were received and how many employee suggestions were acted on so that they can track the performance of their system. I remember some managers taking me aside and explaining why they were interested in SafeStart. They told me, 'We are also interested in doing this because we think it's going to have a huge impact on our quality...and we're hoping it improves safety, as well.' They were already three times safer than their nearest competitor in terms of safety performance in North America, so they didn't have a pressing safety concern—at least not in their opinion. There was certainly no pressure from the regulators. But they had such a strong quality culture that, unofficially, they would tell me that one of the reasons they were doing this was that they firmly believed SafeStart was going to improve quality. They also said that if the powers that be will let them, they would share their quality improvement numbers with us after we complete the implementation. We're just now starting to get a look at some of those numbers, and it would appear that those numbers are indeed improving. So obviously, making fewer errors is going to improve quality. It certainly is going to eliminate accidental equipment damage and downtime."

"They also got a 41 percent decrease in recordable injuries within a year," Gary said. "And even though they didn't have any pressing concerns, that amount of injury reduction got the other division (20,000 employees) interested."

"One of the things that we have not talked about a lot," I said, "is the effect SafeStart concepts can have on customer service. We have all run into customer service people who were in too much of a rush for us, or they were frustrated with something and didn't help us out much. It goes on and on. What kind of effects have you seen in terms of customer service?"

"We did some work with a fairly large airline that was having trouble in a lot of areas including safety, customer service and quality," said Gary. "I thought it was interesting when you told me that their motto was 'We're not happy 'til you're not happy.'"

"It was supposed to be a joke," I quickly interjected.

"Well, in spite of that, this organization had a really neat turnaround," Gary continued. "They got their lost time accidents to zero within six months. We were mainly working with the ground crews who handled and moved baggage. Of course, not making mistakes is a big part of customer service."

"Customer service is also important in the hospitality industry. There's lots of opportunity to see the four states in action: You've got people making mistakes in the kitchen because they're frustrated with the service staff who didn't get the order right, and now they've got to rush, and they're frustrated. You can't control when everybody comes into your restaurant. If a bus arrives, now you've got 50 people who just walked in unexpectedly, and they all want to eat at once, so you're going to be rushing in the kitchen to get all of those appetizers, the salads and the entrees out. There's a lot of coordination that has to be just right for customers to be satisfied. When you think about the hospitality industry in general and kitchens in particular, you have to recognize that there are a lot of hazards that you can't guard: the hot liquid, the hot surfaces, the sharp knives, the slippery floors—there's only so much you can control from an engineering point of view. So you've got a lot of rushing, frustration, fatigue and complacency in the kitchen, and you've got hazards that you can't control. In British Columbia, I found that chefs got injured more frequently than any other occupation. Now, they're not usually getting killed like people in the forest products industry or oil rig workers or the drivers of transport trucks or commercial vehicles, but the frequency of the recordable injuries was the highest I'd seen."

"If you look at health care workers, they're getting hurt at an alarming rate also. They get hurt more frequently than the people out there with the three-foot chainsaws. Hospital employees aren't normally worried about their own safety. After all, they've got seriously injured, sick and dying patients to think about. But they *should* be concerned about their own personal safety because when I started looking at the industries and occupations where there were the most injuries, what I found, surprisingly enough, was that they were not the industries with the most amount of hazardous energy. Instead, they were the industries that had the greatest amount of rushing, frustration, fatigue and complacency. These states were not only leading to more injuries, but they were also having a negative impact on customer service."

"And just one little add-on here," Gary continued. "One of our customers was really struggling in the safety area. But like so many other companies who have safety problems, they also had problems controlling costs, maintaining production levels to ensure on-time delivery of their product and quality issues—issues that all negatively impacted customer service.

The SafeStart training they implemented had a huge impact on each of the issues they faced. Today their logo has the company name, and underneath it says, 'Doing things right.' That's so neat to see that because they wouldn't have been able to put that on there before since they weren't doing things right. But now the organization has turned so far around that they're proud of it, and now they can truthfully say, 'Doing things right.' So, when companies have trained their employees to recognize the state-to-error risk pattern and equipped them with techniques to help them keep their eyes and mind on task, they've not only reduced their injuries, but also improved their customer service."

As Gary talked about that company's turnaround, I remembered that what started my thinking about quality, production, maintenance, and downtime was the manager of a salt mine for a company we worked with. He stayed after the session because he said he wanted to have a word with me. "Sure," I told him. Normally, when a manager tells you they want to have a word with you, it usually has to do with something about how you came very close to crossing the line in terms of political correctness and they want you to apologize to somebody or watch it in the future. So I didn't really know what was coming, but I was concerned.

When everybody was gone, he said to me, "When are you going to take these safety concepts and start applying them to production and maintenance? That's really why I'm here." *Now* I knew why he didn't want to say that in front of everybody else. And I've never forgot what he said. Much of a company's attention is focused, right or wrong, on controlling the most significant costs they have. They want to keep the presses rolling, the pipeline full or the assembly line running. Their attention is on controlling maintenance cost and ensuring downtime has a minimal impact on production. Those are very important considerations from a financial perspective.

So I understood where the manager was coming from. I also understood why he wanted to wait until everybody was gone: He didn't want anybody there thinking that he thought production or maintenance was more important than safety. He was trying to explain to me that financially, the impact on production and quality can be more important than safety.

So if you're a safety professional, you ought to look at your organization and find out what those costs are (damage, downtime) so that you can enlist the support of the production manager, maintenance manager and quality control manager to help you in this effort of creating an inside-out culture. Because when you're looking for culture change, when you're looking to improve things, you need as much help as you can get. You need to start looking at "What's in it for me?" from the point of view of a plant manager, production manager, maintenance manager or quality control manager. Then talk to them about the concepts and techniques and ask them, 'If we could eliminate a lot of the rushing, do you think it will improve quality?' or 'If we could eliminate a lot of the frustration, do you think it could improve our customer service?' If they think it will, then just like the auto manufacturer, they will support it. Getting them to help you to get inside-out thinking through the whole company will be critical, and getting them to see that these concepts and techniques would be a huge help in terms of production, quality, maintenance and customer service will be a powerful incentive for them to get on board.

"The idea we've been promoting here is that inside-out thinking is the best road to a positive safety culture," I said. "We've tried to show how to go about it in terms of integrating these concepts into your workplace. Obviously you need to start with training your employees and getting your leadership group engaged. Then you need to look at your safety management system and the way you run your business, and start rethinking those things from an inside-out point of view so you can reengineer the workplace. Here's the question I'd like to ask you Gary: After 40 years in the safety field, do you think SafeStart is the only road to a positive safety culture, or do you think it's the best road?"

CHAPTER 15—*Managing Complex Change*

"I don't know of any better way to get to a positive safety culture," Gary responded. "There're a lot of organizations that say they have a positive safety culture, but when I get there, I'm not so sure. I don't think implementing SafeStart is the only way of changing culture from the inside out, but it's a much quicker way. I also want to caution people that changing culture isn't easy. So let me go through some things that are required to manage any complex change in an organization, including changing your safety culture."

"The diagram (see Figure 32[8]) illustrates the fact that if you don't have certain things in place, you get into trouble. For example, look at *line one*. There you'll see boxes labeled 'vision,' 'skills,' 'value,' 'resources' and 'action plan.' To the far right you'll see the Greek symbol delta (Δ) that indicates you'll get a change. What I'm saying is if you have articulated a vision, you know where you are and where you want or need to go, and (skipping over to the right), if you have an action plan that is robust, it makes sense to everybody and you execute the action plan, you should see organizational change. But you'll only see it if your organization

---

[8]The Managing Complex Change diagram shown in Figure 32 has a third dimension for each of the five pillars, but for our discussion here, we'll only use the two-dimensional version.

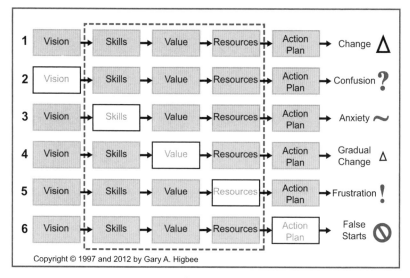

1	Vision →	Skills →	Value →	Resources →	Action Plan →	Change	Δ
2	Vision →	Skills →	Value →	Resources →	Action Plan →	Confusion	?
3	Vision →	Skills →	Value →	Resources →	Action Plan →	Anxiety	~
4	Vision →	Skills →	Value →	Resources →	Action Plan →	Gradual Change	Δ
5	Vision →	Skills →	Value →	Resources →	Action Plan →	Frustration	!
6	Vision →	Skills →	Value →	Resources →	Action Plan →	False Starts	🚫

Copyright © 1997 and 2012 by Gary A. Higbee

Figure 32
Managing Complex Change

has people that are skilled enough to execute the action plan, you provide the resources necessary to execute the action plan as it's designed, and your employees and you and everybody else in the organization sees the value in the endeavor and are willing to participate."

"What's the significance of the dotted blue line that surrounds 'skills,' 'value' and 'resources'?" I asked.

"Oh, yeah, I forgot to mention that," Gary replied. "I drew the dotted blue around them to indicate that these three components are the most troublesome. If the change effort fails, it's likely that one or more of these three was responsible for killing or damaging the effort."

Gary pointed to line two. "Now look down at *line two* where I've taken 'vision' out of the equation. If you don't have a vision that tells you where you are and where you need to go, it doesn't matter how skilled your employees are or how good your action plan is, all you're going to have is confusion."

"So what you're saying," I replied, "is that if you don't have the vision, you might have the skills, you might have the obvious value, you might have the resources, you might have the action plan, but you've got a lot of confusion because people aren't sure where they're going."

"That's right, they need to know the reason for doing what they're doing," Gary said. "Now, on *line three*, you'll notice I've pulled 'skills' out of the mix. So you have a vision, and your employees see the value of the endeavor; they want to do this. You are providing all the resources that are required, and you have this robust action plan put together, but your employees don't have the skills to execute the action plan."

"That's where you get the anxiety. It's just like what we saw at that chemical plant, the one that had lost their positive safety culture, where they had everything but the skills," I said. "It was creating a ton of anxiety."

"Right," Gary said. "You get the same kind of results with people who are asked to do observations and don't have appropriate training on how to do observations. You can really get yourself in trouble if the people don't have the skills."

"Now I'm going to skip line four (we'll come back to it in a minute) and drop down to *line five* where 'resources' is grayed out. What happens here is we have a vision and we have a robust action plan. The employees are charged up, ready to go, and they really know how to do what needs to be done. But for some reason, they're not being given the resources to perform what needs to be done. People get frustrated, in fact, their emotions border on anger. Let me give you an example: Suppose it's an observation process where we have 100 people who are supposed to be doing observations. However, there's always something more important going on and they're never released to do the observations. All they hear is, 'Well, we're behind schedule today, I can't let you do observations, maybe we'll let you do some tomorrow.' Tomorrow comes, but they never get to the point where they do the observations."

"Or you do the observations," I said, "and find some deficiencies. You report the deficiencies and even get them on a work order, but then nothing happens because they don't have the money to fix it, or they *say* they don't have the money to fix it, or whatever. So then what happens is you get the 'What's the use? Nothing ever changes around here' type of frustration."

"Exactly," Gary agreed. "So you can see how each one of these little boxes can affect the other ones."

"I've certainly seen what happens on *line six* (see Figure 33) as well," I said, "where you've got everything going, but you don't have a very good action plan. Like you indicate, you end up with false starts or you've got to re-launch this or do it over again because they didn't have a good enough action plan."

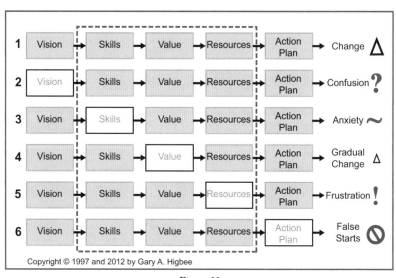

Figure 33
Managing Complex Change

"The one that I think is the most interesting from a safety point of view is *line four*, where 'value' has been removed. This is the part that I think a lot of safety professional miss. They believe that since everybody is motivated not to get hurt—pain is a universal motivator—that everybody sees the value in changing or has enough incentive to change. It's the same assumption I made at the U.S. Postal Service. But, there's a difference between the natural motivation to not want to get hurt and having enough incentive to try to improve—especially when you don't think you have a problem. On line four, the diagram indicates that when people don't see the value in the vision, all you get is very gradual change. What I realized with the post office is that people rarely decide to improve their own safety. After all, most adults think they're safe enough already. You can try to scare them with statistics or warnings of doom and gloom. You can show them gory videos or tell them scary stories, but none of it lasts because they've all come to the conclusion that they're safe enough already. I mean, think about it: If you didn't think you were safe enough, you'd change. If you thought you were driving too fast, you'd slow down. Nobody has ever come into one of our classes and said, 'I hope this will help keep me from getting hurt.' *Nobody* has ever done that. But when we started talking about the safety of their family to those post office employees, we stumbled upon an area of concern or vulnerability. They saw the value in protecting their family, and that motivated them to listen to what we had to say. By involving the family, you can now provide that missing incentive. Of course, the importance of the family isn't new. Many safety professionals already knew that the family was the most important thing to their employees, but they didn't realize that the worker getting hurt wasn't what motivated employees to think about safety—it was the safety of their family."

"Let me just underscore," Gary said, "the importance of employees taking the concepts and techniques home to their family: It's where the value that motivates employees, their incentive, comes from. If you remove that value, you're only going to get gradual change, because adults are not themselves very motivated to work on improving their own personal safety. But they'll be much more motivated to work on improving the safety of their family and their children. When employees learn about the four states, the four errors and the four critical error reduction techniques at work and immediately take them home to teach their own family, they internalize the information and retain it much better (see Figure 34[9]). So even though they think they're 'safe enough' already, by teaching the concepts and techniques to their family they actually improve their own personal safety as well. In other words, they come back to work safer."

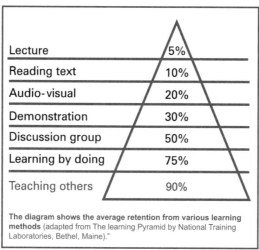

Figure 34
The Learning Pyramid

As Gary and I wound down our conversation—it was getting late and the guy in the boat out on the lake was long gone—I thought about how much ground we had

[9]This graphic or ones similar to it appears in a number of books, scholarly articles and on the Internet. It shows the retention level for seven common instructional methods. The graphic is always attributed to the National Training Laboratories of Bethel, Maine. When contacted recently, the NTL Institute (the current name of the organization) said that they believed that the retention figures were accurate, but "we no longer have—nor can we find—the original research that supports the numbers." We use the retention percentages here because they seem reasonable.

covered. Gary's journey from outside in to inside out had taken him years with a lot of hit and miss along the way. Those who have followed our conversation might be asking, "Why? Why me? Changing the workplace is such a huge job!" Let me see if I can provide a bit of perspective to the issues involved. You asked "Why?" That answer is pretty straightforward: If you want to have a world-class company, you need to focus on reducing injuries 24/7—at work, at home and on the highway. Inside-out thinking using the SafeStart concepts and techniques is the best, quickest way to get there. Your second question was a little more personal: "Why me?" I can't really answer that one for you, but I can sort of point you in the right direction. "Why me?"—You already know a lot about what it takes to keep people safe, especially after you have been introduced to the SafeStart concepts and techniques. You have been given a roadmap for applying those inside-out principles to many aspects of the safety management system. The bottom line, as cliché as it might sound, is it only takes one person to start the ball rolling, one SafeStart champion to make all the difference at your company. Ultimately, it's your decision. Now, one last issue: You're right, changing a company's culture by reengineering the workplace from the inside out isn't for the faint of heart. But as Gary has reminded me on numerous occasions, "The only way you can eat an elephant is one bite at a time." That's good advice with any large task, and changing an organization's thinking from outside in to inside out is enormous. But the payoff is there too—a world class company.

CHAPTER 16—*Moving Foward*

It's now been a number of months since Gary and I finished up our conversation in Whistler. During that time, there have been many significant developments in terms of SafeStart for elementary schools, middle schools, high schools and post-secondary trade schools. This has been primarily centered in Alberta, Canada, but is now spreading into British Columbia and Ontario. At the same time, SafeStart concepts and techniques have been introduced to Canadian athletes at the elite level in hockey, freestyle skiing and basketball.

In the next few years, many more schools and teams will be teaching SafeStart concepts and techniques. The goal is to have SafeStart integrated into the public school system throughout Canada and the USA within the next 10 years. However, it appears that it might spread faster than that in athletics because all the elite athletes are very concerned about getting hurt. They know that one knee, wrist, back or shoulder injury could be the end of the dream. And they know it doesn't matter where or when they get hurt—whether it was in competition, during practice or driving a car—any serious injury could be enough to sideline them. So they're very motivated to learn as much as they can, and so are their coaches. If elite athletes are successful with SafeStart, then there is a good chance it will be accepted fairly quickly by athletes at the house league (recreational) level. So, as mentioned before, it's going to be very interesting to see what's going to happen with schools and athletics in the next 10 years.

What I'm trying to say is that, chances are, SafeStart will eventually get to your community. The real question is whether it will be sooner or later. Kids will eventually learn about rushing, frustration, fatigue and complacency causing or contributing to eyes not on task, mind not on task, line-of-fire, and balance, traction and grip when they are learning to read, write and count. And athletes will learn about the four states, the four critical errors and the four critical error reduction techniques when they are learning, practicing and developing their skills.

So, the question really is, "How long will it take before it gets to your community?" The answer is simple: much longer if you don't get involved and lend a hand. To put it in perspective, just in the time it took you to read this chapter, seven children in the world died from unintentional injuries,[10] and in North America, 48 children were rushed to the emergency department.[11] So when it comes to preventing injuries and fatalities—especially to children—*sooner* is definitely better than later.

We have already given thousands of DVDs and over 10,000 access codes to the online course to youth groups and schools in various communities all over North America. If you'd like to take SafeStart concepts and techniques to a school or youth group in your community, please visit our web site—www.safestart.com—and click on "Corporate Citizenship" under the "Home" tab.

---

[10]Wim Rogmans, "Education and Legislation Are Key to Preventing Child Injuries," Bulletin of the World Health Organization, Volume 87, Number 5, May 2009, < http://www.who.int /bulletin/volumes/87/5/09-050509/en/index.html>, accessed April 9, 2012.

[11]National Safety Council, Injury Facts, 2011 Edition, Itasca, IL, p. 29. The data is from 2008.

CHAPTER 17—*A Truly Positive Safety Culture—24/7*

You can't force everything in safety. You can't enforce eyes on task, mind on task, or balance, traction, grip. You can't prevent people from unintentionally moving into the line-of-fire because they didn't look first. So whether you like it or not, you can't force everything in safety. But you can influence people, and you can improve their natural motivation for not wanting to get hurt. The real motivation comes at the point where they're vulnerable or where they perceive vulnerability, which is with their family. I wouldn't say that implementing SafeStart is the only road to a positive safety culture. But I do think it is the *best* road to a positive safety culture, and it is certainly *a quicker* road to a positive safety culture than going through the traditional five stages to world-class safety. We saw that time and time again.

When you look at a positive safety culture, injury prevention is what everybody really wants. It doesn't matter how disgruntled your workforce is, no one wants anybody to get seriously hurt. So far, these concepts and techniques are the only things that really impact off-the-job safety. One of the most important things that companies misconstrue here, I think, is that while there is a significant risk of dying unintentionally—it's about

1 in 31[12] in terms of lifetime odds—very little of that risk is on the job—lifetime odds of 1 in 801[13]. So when you look at what would actually help people with the overall risk in their lives, it's a mistake or misleading to pretend the workplace is where you should be primarily focused. It's much more important to develop an inside-out culture that focuses on safety 24/7, a culture that begins in the workplace and spreads out to include both the family and the community.

---

[12]National Safety Council, *Injury Facts, 2011 Edition*, Itasca, IL, p. 35. The data is from 2007.

[13]Using 4,829, the worker death rate for 2007 from page 53 of *Injury Facts*, the lifetime odds of dying at work would be 1 in 801.

Index

A

accident/incident investigation, 191–198, 217

accidental equipment damage, 116, 275–276

Aldrich, Mark, 5–6

American Federation of Labor, 12

American National Standards Institute, 13

American Society of Safety Engineers, 10, 69

American Standards Association, 13

analyze close calls (2nd CERT), 113–114
    to fight complacency, 112

B

balance, traction, grip:
    and accident investigation, 195
    and analyzing close calls, 113–114
    and changing what you normally do, 140
    and eyes not on task, 234
    and near miss reporting, 171, 177
    and observation-feedback process, 212
    and redundant devices, 147–148
    and safety-related habits, 104–105, 160
    and self-triggering, 89–90, 100
    and the design stage, 235
    and the difficulty of zero injuries, 170–171
    and the state-to-error risk pattern, 156
    example of, 133–139
    unenforceable, 293

behavior-based safety:
    and ABC theory, 41, 216
    and deliberate risk, 41, 131, 133
    and habit strength, 102
    and human error, 41–46
    and improved awareness, 109–110
    and injury causation, 133

and unsafe condition, 258–259
    observations, 26–27, 211–212, 254
    slow process, 255–256
    sustaining the process, 47–48
    three stages (Larry), 31

blame game, 191, 193, 217
    see also "self" area; misunderstanding of

Board of Certified Safety Professionals, 13

Bureau of Mines, 7

C

Carder, Brooks, 68

chain reaction (accident causation), 132, 178, 189, 192, 270

circadian rhythms, 228

Coastal Training Technologies, 119

complacency, 78, 87, 148–149
    and compliance training, 199–201, 217–218
    and decision-making, 133–135, 147
    and distracted driving, 142–143
    and emergency preparedness, 204–206
    and job safety analysis, 183
    and looking at others, 112–114, 143, 216
    and mind not on task, 101, 114–115, 147
    and observation-feedback process, 212–213
    and quality problems, 276–278
    and redundant devices, 147–148
    and risk assessment, 186–187, 235
    and safety-related habits, 114, 181
    and self-triggering, 101
    and the state-to-error risk pattern, 86–87, 156
    and trusting things to memory, 141–142
    discovery, 78
    examples of, 102–103, 112–113, 133–139, 187–189, 204–205, 233–235
    level of, 143, 146–147
    like oxygen in the fire triangle, 231–232
    prevalence, 186

F

Fair Labor Standards Act, 13
fatigue:
    and accident investigation, 193
    and analyzing close calls, 114
    and complacency, 140
    and compliance training, 200–201,
        245
    and decision-making, 133, 136,
        139–140, 147, 149
    and job safety analysis, 183, 189
    and mentoring, 242
    and observation-feedback process,
        212–214, 258
    and quality problems, 276, 278
    and recognizing change, 140–141
    and redundant devices, 148
    and risk assessment, 217
    and self-triggering, 90, 100, 140
    and the design process, 235
    and the "self" area, 178
    and the state-to-error risk pattern,
        86–87, 156
    and workplace procedures, 219–221,
        227–230
    signs of, 144
    versus physical hazards, 159, 208,
        237
fatigue weave, 144
Federal Coal Mine Safety Act, 13
Federal Employers Liability Act, 7
Federal Metal and Nonmetallic Mine
    Safety Act, 13
Five Stages to World Class Safety
    (diagram), 17–33, 253–255
    Stage 1, 18–19, 253
    Stage 2, 19–22, 254
    Stage 3, 22–27, 254
    Stage 4, 27–29, 255
    Stage 5, 29–30, 255
frustration:
    and accident investigation, 193
    and analyzing close calls, 114
    and compliance training, 200–201,
        245
    and customer service, 280
    and decision-making, 133, 136,
        139–140, 147, 149
    and job safety analysis, 183, 189

    and managing complex change, 286
    and mentoring, 242
    and morale, 271
    and observation-feedback process,
        212–214, 258
    and quality problems, 276, 278
    and recognizing change, 140–141
    and redundant devices, 148
    and risk assessment, 217
    and rushing cycle, 271–272
    and self-triggering, 90, 100, 140
    and the design process, 235
    and the "self" area, 178
    and the state-to-error risk pattern,
        86–87, 156
    examples of, 78–81, 81–83,
        176–177, 207–208, 208–209,
        219–223, 278
    versus physical hazards, 159, 208,
        237

G

Gompers, Samuel, 12

H

hazard recognition, 184–186
housekeeping, 275
human error:
    and accidents, 191
    and behavior-based safety, 41–43
    and ergonomists, 38
    and SafeStart, 4, 177
    and safety professionals, 39–40
    and "Utopia," 48
    and observation-feedback process,
        258
    in popular thought, 49–50
    versus the safety system, 48–49
Hurt at Home (DVD), 125
    and enhancing accident/incident
        investigation, 194–195, 217

I

"I'm safe enough," 73–76, 107, 115–116,
    123–124, 261, 287–288

O

Occupational Safety and Health
Administration:
see OSHA

observation-feedback process, 211–218
and SafeStart, 258, 260
and total recordables, 54

OSHA:
and behavior-based safety, 111
and ergonomics, 38
and Stage 2 (Gary), 19
early years, 14, 158
inspections, 21
Region 2, 111

outside-in thinking (extrinsic control):
and personal protective equipment,
207
and railroad industry, 7, 259
and regulations, 8
and workers' compensation laws,
9, 13
limitations of, 184

P

paradigm shift, 3–4, 151, 157–159

peer-to-peer observations:
at Stage 4 (Gary) scheduled, 27–28,
255
at Stage 5 (Gary) in real time, 29,
36, 255
challenges of, 47–48, 255–256
enhancing observations, 212
examples of, 29–30, 258–260

Perkins, Frances, 13

permanent partial disability, 12

positive safety culture, 252, 259,
263–264, 267–268, 280, 283, 285,
293–294

preventive maintenance (PM), 269

productivity, 116, 219, 221, 237,
241–242, 271, 273

Q

quality, 219, 221, 227, 230–231, 237,
241–242, 271, 273, 275–280

R

reengineering, 149, 232, 238, 289

risk assessment, 181–184, 186–190, 199,
216–217

rushing:
and accident investigation, 193, 195
and analyzing close calls, 114
and complacency, 140
and compliance training, 200–201,
245
and decision-making, 133, 136,
139–140, 147, 149
and frustration cycle, 271–272
and job safety analysis, 183, 189
and level of complacency, 135–136
and mentoring, 242
and observation-feedback process,
212–214, 258
and quality problems, 276, 278, 280
and recognizing change, 140–141
and redundant devices, 148
and risk assessment, 217
and self-triggering, 89–90, 100, 140
and the design process, 235
and the "self" area, 178
and the state-to-error risk pattern,
86–87, 156
examples of, 112–113, 160–162,
176–177, 208–209, 219–223,
223–225, 225–227
discovery, 67–71
"legitimate" rushing, 223
prevalence, 129
versus physical hazards, 159, 208,
237

rushing-frustration cycle, 271–272

S

SafeStart errors:
see eyes not on task
see mind not on task
see line-of-fire
see balance, traction, grip

SafeStart Home series (DVDs), 123, 125,
131, 194, 262–263, 265, 272

SafeStart states:
see rushing

About The Authors

Larry Wilson started out as a behavior–based safety (BBS) consultant when he was 28. By the time he was the age of most BBS consultants and trainers, he had more practical experience— more days, more companies, more industries—than most would get in their entire consulting careers. He also worked in some of the harshest environments, from hand–logging on the Pacific Coast to oil drilling beyond the Arctic Circle. He is the author and creator of SafeStart, a training program that focuses on reducing human error. SafeStart is currently being used by more than two million people in 40 countries.

Gary A. Higbee, EMBA, CSP, has been a plant, division or corporate safety director for a number of Fortune 500 companies, including Maytag, Budd Company, Electrolux and Deere & Company. He is a Certified Safety Professional who has received the Safety Professional of the Year award from the ASSE Region IV. He also received the Distinguished Service for Safety Award—the highest award given to an individual—from the National Safety Council.